FLORISSANT BUTTERFLIES

A Guide to the Fossil and Present-Day Species of Central Colorado

FRONTISPIECE. The 35-million-year-old holotype specimen of *Vanessa amerindica* found in the Florissant shales, and described as a new fossil species of butterfly in 1989 by Jacqueline Y. Miller and F. Martin Brown.

Florissant Butterflies

A Guide to the Fossil and Present-Day Species of Central Colorado

Thomas C. Emmel, Marc C. Minno, and Boyce A. Drummond
Department of Zoology, University of Florida,
Gainesville, Florida, and
Pikes Peak Research Station, Florissant, Colorado

Stanford University Press
Stanford, California
1992

Stanford University Press
Stanford, California

Printed and bound by CPI Group (UK) Ltd,
Croydon, CR0 4YY

The photo on the front of the book is of
a mating pair of Phoebus Parnassians
(*Parnassius phoebus sayii* W. H. Edwards)
on a Florissant mountainside.
(Photo: B. A. Drummond)

ISBN 978-0-8047-2018-2
ISBN 978-0-8047-1938-4

Emmel, Thomas C.
Florissant butterflies: a guide to the fossil and present-day species of central Colorado/Thomas C. Emmel, Marc C. Minno, Boyce A. Drummond.
p. cm.
Includes bibliographical references and indexes.
ISBN 0-8047-1938-1 (alk. paper), ISBN 0-8047-2018-5 (pbk.)
1. Butterflies--Colorado--Florissant Region. 2. Butterflies. Fossil--Colorado--Florissant Region. 3. Butterflies--Colorado--Florissant Region--Identification. 4. Butterflies, Fossil--Colorado--Florissant Region-Identification. I. Minno, Marc C. II. Drummond, Boyce A. III. Title.
QL551.C6E45 1992
595.78'9'09788--dc20 91-13262
CIP

PREFACE

Butterflies are an attractive and conspicuous element of our natural environment. They also appear as delicate rare fossils from millions of years ago, and the Florissant Fossil Beds National Monument in central Colorado is world-famous for its butterfly fossils as well as thousands of other prehistoric insect fossils. Today in the Florissant region, near Pikes Peak, nearly 100 species of living butterflies wend their way over flower-spangled meadows and pine-crested ridges throughout the spring, summer, and fall. Many of these species have unusual life histories. For example, the larvae (caterpillars) of a species of *Oeneis* take two years to mature, and many lycaenids develop close associations with ants. The adults of many species live only a few days, whereas those of some butterflies survive for months in winter hibernation. Regardless of the specifics of the biology, butterflies never fail to elicit a response of wonder and admiration in the observer. Thus, this book was written not only to summarize the scientific details and natural history of Florissant butterflies, past and present, but also to introduce the visitor in the Pikes Peak region to this fascinating group of insects. We hope you will enjoy reading the book and using it to identify your sightings. Perhaps it will stimulate you to make new observations about the butterfly fauna of the central Colorado Rockies, one of the richest of the land, and to add to our knowledge of these fascinating and beautiful creatures.

* * * *

Mr. and Mrs. Roger A. Sanborn, Directors of the Colorado Outdoor Education Center and its associated programs, have encouraged natural history studies and research on the butterfly fauna of this area since the senior author initiated this project in 1960. Their founding of the Pikes Peak Research Station and its programs, especially the Lepidoptera Week and, Biology of Butterflies Workshops, brought the authors together on the present publication, and we are deeply indebted to them for their constant support and strong interest.

The authors would also like to thank James L. Nation, Jr., for his many contributions during the last ten years of field work in the Florissant region, as well as for help in the preparation of the color plates; Christine Eliazar for her extensive contributions in preparing many drafts of the manuscript; Maria Minno and Peter J. Eliazar for reviewing and editing the manuscript drafts; Dr. Charles V. Covell, Jr., and hundreds of other lepidopterists who have participated in the Biology of Butterflies Workshops held annually at the Colorado Outdoor Education Center and Pikes Peak Research Station; and Dr. John F. Emmel and many student research assistants from the University of Florida who over the last 25 years have contributed their time and specimens in helping to determine the life histories and pursue the field work reported in this book.

The Superintendent and his National Park Service staff at Florissant Fossil Beds National Monument have been most generous in assistance, and in lending specimens for study and photography. Dr. F. Martin Brown, dean of the Rocky Mountain lepidopterists and principal author of the outstanding book on the Colorado butterfly fauna (*Colorado*

Butterflies, 1955), is also the world's authority on the fossil insects of Florissant and has generously shared his knowledge, time, literature, and photographs of Florissant fossils with us. Dr. Reinhard A. Wobus (Department of Geology, Williams College) critically reviewed the geological sections and fossil discussions, and materially improved both the accuracy and currency of these topics as of information available through July 1991. Miss Daryl Harrison, Biological Illustrator for the Department of Zoology at the University of Florida, kindly prepared the map of the Florissant region, and Dr. Ronald G. Wolff of that department generously offered the use of his photographic facilities for the preparation of the color plates.

T. C. E.
M. C. M.
B. A. D.

CONTENTS

(Twenty pages of color plates follow p. 54)

FIGURE 1. Lepidopterist enthusiasts explore mixed dry and wet meadow habitat near Dome Rock at the southern end of the Florissant Valley.

FLORISSANT BUTTERFLIES

A Guide to the Fossil and Present-Day Species of Central Colorado

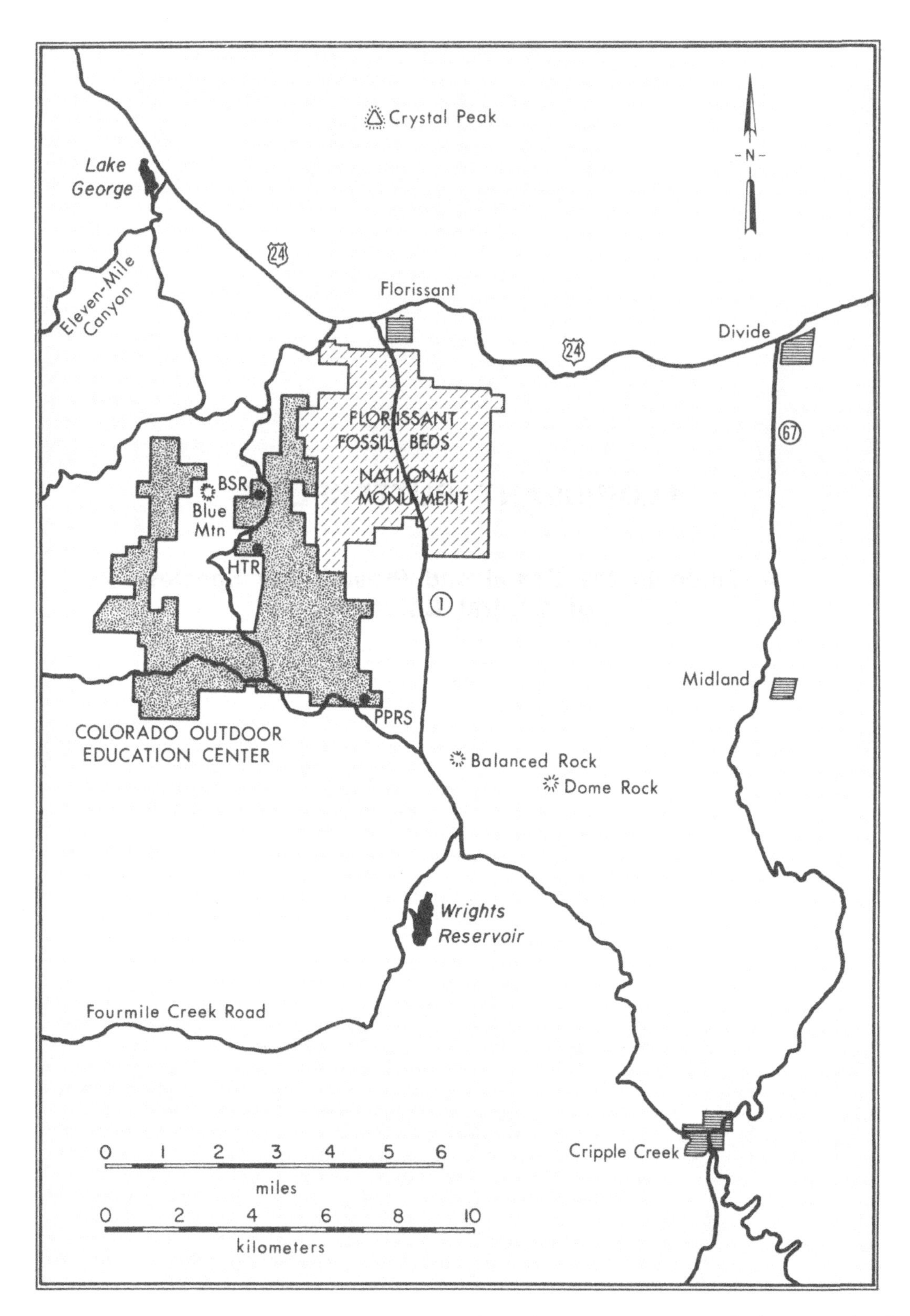

FIGURE 2. The area covered by this faunal work on the butterflies of the Florissant region is shown on this map of the Florissant Valley and environs. The town of Florissant is located approximately 35 miles by road (Highway 24) west of Colorado Springs. The Florissant region extends from Crystal Peak to about 4 miles south of Cripple Creek, and from Highway 67 west to Eleven-Mile Canyon and Four-Mile Creek Road. BSR = Big Spring Ranch headquarters. HTR = High Trails Ranch headquarters. PPRS = Pikes Peak Research Station. The heavily shaded area incorporating these three centers shows the 6,000 acres occupied by the Colorado Outdoor Education Center, where most of the field work reported in this publication has been carried out. Topographic features such as Blue Mountain, Crystal Peak, Dome Rock, Balanced Rock, Wrights Reservoir, and other landmarks can be found on the U.S. Geological Survey topographic maps for this area.

INTRODUCTION

To find fossils as perfectly preserved as the minute and fragile insects at Florissant is one of the major wonders of nature. To find that as many as 20 of these insect specimens are delicate-winged butterflies is even more remarkable. These fossils have endured through 35 million years of geological activity, uplift, erosion, and climatic change to fill us with awe today.

But surrounding us in central Colorado and the Florissant Valley are many wondrous living species as well. The region supports nearly 100 modern butterfly species, which wend their way over meadow and through forests from early spring to late fall. We do not have to come to Florissant by horseback or on the old stage road from Colorado Springs to South Park, or ascend the old railroad grade as in the 1870's to early 1900's, when the first fossil collectors found this wealth of insect material following the U.S. Geological Survey expedition led by Dr. Hayden. Instead, we can arrive by car in less than an hour from Colorado Springs and begin to experience a contemporary exploration of the rich natural history, past and present, of the Pikes Peak region in and around the Florissant Valley (Figure 2).

The climate is cool and dry today, compared with the subtropical and warm-temperate climate presumed by some geologists to be present 35 million years ago. Pines and aspens edge the grassy valley meadows, the latter covered by summer wildflowers, while long ago in the Oligocene Epoch, when this land lay at a lower elevation, giant redwoods lined the streamsides and chaparral covered the higher ridges. Come with us now and travel back in time some 35 million years to when butterflies filled the air above a beautiful shallow lake, and great nearby volcanoes began erupting and entrapping these insects. Then we will jump forward to the present, and look at the rich diversity of modern habitats and the 97 butterfly species that can be found today in the Pikes Peak region of the southern Rockies.

FIGURE 3. The west slope of Pikes Peak, showing timberline and the lower broad expanse of montane forest across this nearly three-mile-high mountain in central Colorado.

FOSSIL INSECTS AND THE GEOLOGICAL RECORD OF EARTH

The insects, with their six legs, and the spiders and centipedes and other arthropod relatives are thought to have evolved from a common, unknown ancestor early in the Devonian Period of the Paleozoic Era, about 400 million years ago. The evolution of insect wings occurred later, probably during Mississippian times in the Carboniferous Period, about 350 million years ago (Kaesler 1987). Some groups such as cockroaches evolved the ability to fold the wings next to the body, whereas other groups such as dragonflies, not needing to run or walk on the ground, kept their wings extended. The Cretaceous Period of the Mesozoic Era, roughly the period 144 to 66 million years ago, marked the first general appearance of the butterflies and moths of the order Lepidoptera; we have a few adults, larvae, wing scales, and even a fossil egg from the Cretaceous rocks (Gall and Tiffney 1983).

This new group of "scale-winged" insects (Lepidoptera) evolved simultaneously with the diversification of the great modern group of plants called angiosperms: the flowering plants. Today, we find that butterflies and moths are important pollinating insects, and it is likely that their first representatives visited the first flowering plants for nectar and pollen foods. We presume, too, that the larvae of the first Lepidoptera fed primarily on the leaves and stems of flowering plants, just as they do today. Thus, it is not surprising that the order Lepidoptera should first appear in the fossil record at the time of the major adaptive diversification of their hosts. Today, with over 200,000 named species, the flowering plants are the dominant plant group in the world, and the modern lepidopteran species match or exceed their hosts in number of species.

How did the Lepidoptera diversify in the millennia following their origin, 100 million years ago? It is believed (Shields 1976, 1988) that the Lepidoptera, along with their closest insect relatives, the caddisflies (order Trichoptera), originated in the Mesozoic from scorpionfly-like insects (order Mecoptera). These earliest Lepidoptera arose in what is now Queensland in northeastern Australia, where the distributions of the first fossil Trichoptera genera, the most primitive modern moth genus (*Agathiphaga*), and the most primitive living butterfly genus (*Euschemon*, a Pyrginae skipper) all overlap (Shields 1988). Yet this view derives only from indirect evidence. The oldest lepidopteran fossil known is *Archaeolepis*, a moth that was recovered from rocks at Dorset, England, and which dates from the earliest part of the Jurassic Period (208 million years ago).

The oldest undoubted butterfly fossils, as distinct from moths, are four specimens collected in the Green River Shale of Colorado, at Raydome in Rio Blanco County, and dated at 48 million years, at the middle of the Eocene Epoch. Three of these have been described as two primitive papilionids (swallowtails) and one riodinid (metalmark) species; a fourth specimen, apparently a satyrine nymphalid, is not yet described (Durden and Rose 1978). The next earliest butterfly fossil dates from the boundary of the Eocene and Oligocene Epochs (about 37 million years ago) in the Tertiary Period of the Cenozoic Era. It is a lycaenid larva and was found in Baltic amber. Two papilionids from the same time frame were also found in Baltic amber. But the richest fossil butterfly fauna in the entire world occurs in the Florissant shales at 36 to 34 million years, and suggests (with representatives of the butterfly families Pieridae, Libytheidae, and Nymphalidae being present in relatively modern-looking species) that all the basic families of butterflies had already evolved millions of years earlier, probably by the end of the Cretaceous Period about 66 million years ago, when some of the continents were in much closer geographic proximity (see also Shields 1987).

By the end of the Oligocene and the start of the Miocene Epoch (about 24 million years ago), butterfly fossils from all the modern families appear in rocks worldwide. These sources include calcareous marls of gypsum quarries in Aix-en-Provence, France (1 hesperiid, 1 papilionid, 1 pierid, 3 satyrid, and 1 lycaenid species); beds of freshwater sands, clays, and limestones at Gurnet Bay on the Isle of Wight (1 nymphalid); cemented marine sand near Szczecin, Poland (1 unplaced adult), lignite beds in Bavaria, Germany (1 hesperiid); marls of lacustrine beds at Radoboj, Yugoslavia (1 pierid, 2 nymphalids); seacoast limestones at Gabbro, Italy (1 lycaenid, 1 papilionid), deposits at Stavropol, North Caucasus (2 nymphalids); and paper-coal deposits (extremely fine shale) at Randecker Maar in southwestern Germany (1 pierid) (Shields 1976). The famed anthropologist Louis Leakey found undetermined Lepidoptera larval and pupal fossils in Miocene sandstone and clay beds from two islands in Lake Victoria, Kenya (Leakey 1953).

During the Quaternary Period's Pleistocene Epoch (1.6 million to 10,000 years ago), butterfly fossils appear in lake deposits at Shiobara, Japan (1 papilionid and 2 hesperiids); in amber in East Africa (1 pierid), and in lacustrine beds in Re, Italy (1 lycaenid and 1 unplaced pupa). Shields (1976) also reports 1 fossil nymphalid and 1 unplaceable adult butterfly in amber in Tanzania, from the Recent Epoch, dated from 2,000 to 3,000 years ago.

This record of about 44 known and published butterfly fossils is the extent of our current knowledge about the evolution of butterflies through the fossil record, across 48 million years. Obviously, much scientific work and further exploration for fossils of Lepidoptera remain to be done, and the Oligocene shales at Florissant Fossil Beds National Monument are among the most important sites in the world for future study.

Geologic Time Scale*

Era	Period	Epoch	Start of Period (Years before Present)
PRECAMBRIAN			4.6 billion
PALEOZOIC	Cambrian to Permian		570 million
MESOZOIC	Triassic		245 million
	Jurassic		208 million
	Cretaceous		144 million
CENOZOIC	Tertiary	Paleocene	66 million
		Eocene	58 million
		Oligocene	36 million
		Miocene	24 million
		Pliocene	5 million
	Quaternary	Pleistocene	1.6 million
		Recent	10 thousand

*All dates in this table and in the text are based on the current geological time scale dating of eras, periods, and epochs approved by the Geological Society of America (1983 decade).

THE FOSSIL BUTTERFLIES OF FLORISSANT

For more than a century since the Hayden Survey in 1873, the ancient lakebed sediments at Florissant have been renowned around the world for their thousands of fossil insects and plants. Even delicately preserved birds as well as fishes have been found in these shales. The age of the lakebeds, which cover about 15 square miles, has been variously assigned to the Oligocene (36.6-23.7 million years ago) or Miocene (23.7-5.3 million years ago) epochs of the Age of Mammals (Cenozoic Era), but today they are generally acknowledged to date between 34 and 36 million years ago in the early Oligocene. Beneath the lakebeds lies an ancient Precambrian basement-rock complex more than a billion years old (Durden 1966).

About 36 million years ago, an ancient stream flowed south through the Florissant Basin, which then lay at an elevation of about 3,000 feet (900 meters) above sea level. However, work being done currently by Kathryn M. Gregory at Florissant suggests that the elevation may have been over 2,000 meters, similar to the elevation today. Explosive volcanic eruptions in the Sawatch Range, 50 miles (80 km) to the west, poured great volcanic ash flows to the east, including the Florissant region, forming the Wall Mountain Tuff deposits that now underlie the Florissant shales. About 35 million years ago, substantial volcanic eruptions of a different kind began to take place in the nearby Thirtynine Mile volcanic field, centered near the village of Guffey 17 miles (28 km) to the southwest. The eastward flow of lava and mud from Guffey blocked the stream drainage in the Florissant Basin, forming ancient Lake Florissant north of this new dam. Even huge *Sequoia* trunks were entombed and silicified by the hot mud flows at this time, still standing upright. (They would remain until the turn of the century, when the first settlers in the Florissant Basin discovered trunks of 30 feet or more in height in areas where the surrounding soil had been removed. Unfortunately, some were cut into pieces for souvenirs and for exhibition elsewhere.) The climate around this lake (which lasted thousands of years) has been described as warm, moist, and almost subtropical, judging from the plants present at that time (MacGinitie 1953), although Gregory (unpublished) has estimated recently that the paleotemperatures and rainfall in the area actually may have been similar to today's annual ranges.

Eruptions of the Thirtynine Mile volcanoes occurred repeatedly from 35 to 34 million years ago, sending great clouds of fine ash and dust over Lake Florissant. Insects, including butterflies, died in the air or water and were carried to the lake bottom by the settling ash. There, the Florissant shales began to form, preserving these delicate fossils. By about 34 million years ago, volcanic mudflows called lahars had covered the lake as the Thirtynine Mile volcanoes continued to erupt. These mudflows hardened and protected the shale beds from erosive forces.

Aside from the recent disturbance by man and natural erosion by streams, the fossil shales have remained intact for 34 million years, surviving even a great geological uplift that raised them to their present elevation of about 8,400 feet (2,520 m) above sea level. Today, the shale beds are close to the surface after erosion of overlying material. The climate has cooled, and a relatively sparse Ponderosa Pine forest and grassy meadows have replaced the moist forests of the Oligocene lakeside habitats.

As mentioned above, the extensive basin of ancient Lake Florissant was filled in slowly by the fine airborne ash erupting from the Thirtynine Mile volcanoes to the southwest of the lake. This fine ash formed thin layers of sediment that allowed the preservation of tiny insect specimens (Scudder 1890) as well as numerous plant remains. With a warm temperate climate, the streamside forests contained such trees as redwoods, maples, oaks, elms, sycamores, and poplars, as we see in the abundant plant fossils of the Florissant shales. The higher terrain supported pines and evergreen oaks, growing in a chaparral plant formation. To find similar modern-day vegetation, one would have to travel south to at least the northern Sierra Madre of northeastern Mexico (Durden 1966). MacGinitie (1953) estimated that the mean annual temperature at that time in the Florissant region was at least 65 degrees Fahrenheit, with an average temperature in the warmest months of approximately 80 degrees F (18°C). An annual rainfall of around 20 inches (51 cm) probably fell principally as summer rains. These estimates of the climate are being revised currently by Gregory's investigations, however (Gregory, unpublished).

Since it is believed that Lake Florissant existed for a period of about 2,500 to 5,000 years (as determined from the rate of layer deposition), it is possible that not all the fossil butterflies existed there at the same time. The species we currently know of may have been separated from each other by several thousand years of geologic time. Certainly, the subtropical and chaparral types of vegetation during the Oligocene Period around ancient Lake Florissant are quite different from the modern flora, which is composed of coniferous forest and meadow habitats (Emmel 1964). It is thus not surprising also that the modern Florissant butterfly fauna displays very little close relationship to the fossil butterfly genera that occur in the lake deposits.

A total of 12 species of butterflies has been described from the Florissant Fossil Beds. These include species from three different families: the whites or Pieridae (two species), the snout butterflies or Libytheidae (two species), and the brush-footed butterflies or Nymphalidae (eight species) (Cockerell 1922; Shields 1985a, b; Miller and Brown 1989). The two butterflies in the Pieridae and seven of the Nymphalidae species appear to have a neotropical relationship, being especially close to modern northern neotropical species. The eighth nymphalid species, *Vanessa amerindica,* is the first described taxon from the Florissant shale deposits to be closely aligned with still-living species from the Holarctic (temperate Europe, Asia, and North America) and the Indo-Malayan (southern Asia) geographic regions. The snout butterflies resemble modern libytheids from southeastern Asia, and perhaps an African species.

The butterfly species that have been named from the Florissant Fossil Beds are as follows:

Family Pieridae, Subfamily Pierinae

1. *Stolopsyche libytheoides* Scudder
   GREAT PALPI WHITE

This pierid butterfly was named by Samuel H. Scudder in 1889 (*Eighth Annual Report, U.S. Geological Survey,* pp. 467-468, Pl. LIII, Figs. 1-3). The specific name *libytheoides* is in allusion to the great length of the palpi (sticking out from the front of the head on either side of the tongue), which resembles the palpi condition found in the snout butterflies of the family Libytheidae. The species is represented by a single fossil, the type specimen (from Florissant), lying on its side. The wings of this butterfly are badly preserved and the outer margin broken off, though parts of the body sufficient for generic determination can be made out with difficulty. The probable length of the forewing was 25 mm, while the length of the palpi is 4.5 mm.

The butterfly appears to have been about the size of the modern Cabbage White, *Pieris rapae* Linnaeus, but with a relatively smaller head, larger thorax, and much larger palpi. The wings were apparently white or at least very pale, and there were no interior dark markings on the underside, unlike the present *Pieris rapae.* Instead, there were grayish edgings to the veins and front margin of the hindwings, the only wings well preserved in the type specimen. It was allied with the Pierinae "whites" and, except for its huge labial palpi, was closer to *Pieris* than any other living genus.

2. *Oligodonta florissantensis* Brown
FLORISSANT PIERID

Described by F. Martin Brown in 1976 (*Bulletin of the Allyn Museum*, No. 37: 1-4, Figs. 1-3) from a very fine fossil, this specimen was originally recovered from the trench on the Singer Ranch at Florissant and passed (with the Ranch property) to the National Park Service when the federal government established Florissant Fossil Beds National Monument in 1969.

While the general shape and position of this pierid make it look rather like a second specimen of the libytheid butterfly *Barbarothea florissanti* Scudder, Brown noted that the palpus (a brush-like structure projecting forward from the front of the head on either side of the head in butterflies) is proportionately smaller and not at all like the palpus of a snout butterfly in the family Libytheidae. Also, the venation of the wings told Brown that he had a pierid butterfly. He believes its closest living relatives are the species of the neotropical montane genus *Catasticta.* No pattern features can be recognized on the wings. The butterfly was fossilized as a fresh specimen (the preservation is so good along the margin of the hindwing that the rows of individual scales can be seen under the microscope), lying on its side with its wings folded over its thorax. The tongue (proboscis) is coiled just below the palpi, and the position of the eye is plainly evident. The radial length of the forewing is about 26.5 mm, and that of the hindwing about 22 mm. The generic name *Oligodonta* alludes to the age of the specimen, late Oligocene, and to the blunt tooth on the margin of the forewing.

Family Libytheidae, Subfamily Libytheinae

3. *Prolibythea vagabunda* Scudder
VAGABOND SNOUT

The first of the Libytheidae found at Florissant, this snout butterfly is most closely related to the modern *Libythea celtis* Fuessly, a widespread European and Asian species of this small family. In fact, Shields (1985b) even places *vagabunda* and the following species, *Barbarothea florissanti*, in the modern Old World genus *Libythea.* Samuel H. Scudder originally described both the genus and the species in 1889 (*Eighth Annual Report, U.S. Geological Survey*, pp. 461-467, Pl. LIII, Figs. 4-9), based on a single specimen collected at Florissant. The fossil specimen is lying on its back, with the underside of the body and some of the various appendages in view. The wings of one side are preserved, but they overlap almost completely. Thus the markings are obscured, but some of the color pattern and the wing veins can be determined. The front pair of legs in this libytheid and the shape and size of the abdomen indicate that the specimen was a female. Besides the affinities to the modern *Libythea celtis* mentioned by Shields (1985b), Scudder (1889) felt that this fossil species was close to the African *Libythea labdaca* Westwood, the only modern species agreeing with the fossil in size. *Prolibythea vagabunda* is the largest snout butterfly known, fossil or living, with a body and palpi length of 31.5 mm and a probable wing expanse of 63 mm (length of restored forewing is 29.5 mm). The specific name is an allusion "to the far-away immediate allies of the fossil, and its relation to a vagabond type" (Scudder 1889).

All of the modern Libytheidae (12 species worldwide) with known life histories feed as caterpillars on hackberry trees (*Celtis* species) (Shields 1989a). Among the fossil plants in the Florissant shales are perfectly preserved leaves and flowers of *Celtis maccoshi* Lesquereux. As Scudder (1889) notes, "It is, therefore, highly probable that *Prolibythea vagabunda* fed on *Celtis maccoshi*" some 35 million years ago.

4. *Barbarothea florissanti* Scudder
FLORISSANT SNOUT BUTTERFLY

This snout butterfly (both genus and species) was named by Samuel H. Scudder in 1892 (*Bulletin of the United States Geological Survey*, 92: 20-24, Pl. 3, Figs. 1-5) from a single specimen. The stone on which it was preserved (lying on its side) is irregular in surface, leaving some points obscure, but overall the butterfly is remarkably well preserved, including the wing venation, the antennae, the extended palpi in front of the head, the drooping tongue (proboscis), and part of the legs. The length of the forewing is 28.5 mm. A faint pattern is visible on this butterfly's ventral surface. The specimen was collected by S. H. Long and resides at the Philadelphia Academy of Natural Sciences. Shields (1985a) believes this fossil species is most closely related to the living snout butterfly, *Libythea geoffroyi* Godart, which occurs on the Indo-China peninsula (Burma, Thailand) throughout the Southeast Asian islands to the edge of northern Australia (Shields 1985b).

Family Nymphalidae, Subfamily Nymphalinae

5. *Prodryas persephone* Scudder
PERSEPHONE NYMPHALID

This nymphalid was described by Samuel H. Scudder in 1878 (*Bulletin of the United States Geological and Geographical Survey of the Territories*, 4(2): 520-526, no figs.), but not figured until later (Scudder 1889, Pl. LII, Figs. 1-10). The single specimen was found at Florissant (probably at Costello's Ranch) by Mrs. Charlotte Hill, "in a wonderful state of preservation, the wings expanded as if in readiness for the [collection display] cabinet and absolutely perfect, with the exception of the tail of the right hind wing." The dorsal surface is uppermost in the fossil, with perfect preservation of the thorax and abdomen but only indications of the legs beneath the wings. The head is twisted to one side, and exhibits the palpi as well as both antennae from the side. The wings appear to be dark brown, with pale markings (which may have been bright colored in life). The most prominent markings on the forewing are five unequal pale spots that cross the wing in a straight line from the lower outer corner to the upper margin about two-thirds of the way out from the body. The hindwings have a very large pale spot covering the entire upper outer quadrant of each wing. Even the individual scales on the outer half of the forewing can be distinguished. The forewing length is 24.5 mm. The type specimen is in the Museum of Comparative Zoology at Harvard University. This butterfly is closest in appearance to the modern neotropical genus *Hypanartia*, found throughout Central and South America and the Caribbean islands. This species and the next two were named by Scudder after figures of the Underworld in Greek mythology. Persephone, maiden of the Spring and a daughter of Zeus, was carried away from earth by Pluto (Hades) to become his Queen of the Underworld.

6. *Lithopsyche styx* Scudder
RIVER STYX NYMPHALID

Lithopsyche styx was described by Scudder (1889) from a single specimen collected some years before 1889 at Florissant by Mr. Israel C. Russell, in whose collection it resided at the time of Scudder's description (S. H. Scudder, 1889, *Eighth Annual Report, U.S. Geological Survey,* pp. 452-457, Pl. LII, Figs. 11, 16, and 17). This species appears to be allied to the extinct genus *Jupiteria* and the modern neotropical genus *Hypanartia.* The wings on one side are imperfectly preserved; the rest of the original butterfly is gone. Scudder was able to restore the missing portions by studying the character of the existing veins and wing margins, and he estimated the total length of the forewing at 27 mm (the existing forewing fragment is 22 mm long, and the hindwing is 19-20 mm in breadth). While the two wings are almost completely overlapped in the fossil, Scudder was able to separate the markings of the two wings. The ground color of both wings was dark with lighter markings. On the forewing, these markings are distributed in moderate sized patches mostly in the center of the upper half of the wing. On the hindwing, these markings were clustered into irregular mottling, mostly in the lower half of the outer portion of the wing. The Styx was the mythological river in Hell by which the gods swore unbreakable oaths.

7. *Jupiteria charon* Scudder
CHARON'S NYMPHALID

Described by Samuel H. Scudder in his major 1889 paper on "The Fossil Butterflies of Florissant" (*Eighth Annual Report, U.S. Geological Survey*, pp. 448-452, Pl. LII, Figs. 14-15), this nymphalid butterfly is allied to the extinct genus *Prodryas* and the living worldwide "Buckeye" genus *Junonia* (=*Precis*). It is represented by a single specimen from Florissant that was in the collection of R. D. Lacoe of Pittston, Pennsylvania, at the time Scudder described it. The type specimen was probably collected at Fossil Stump Hill (F. M. Brown, *in litt.*) prior to 1889 and it is presently in the United States National Museum of Natural History. The forewings are triangular, with a somewhat pointed tip. The hindwings are subcircular, longer than broad with a slightly concave border and venation much as in the modern genus *Eunica*, a tropical and subtropical nymphalid group. But other wing veins are close to those of *Junonia.* Most of the wing markings, unfortunately, are obscure and are in the region where the wings overlap on this fossil. No significant appendage portions can be seen. The forewing length is estimated to be 27 mm, from the 24 mm fragment in the fossil. In Greek mythology, Charon was the aged boatman who ferried the souls of the dead across the waters of the Underworld.

8. *Nymphalites obscurum* Scudder
OBSCURE NYMPHALID

In 1889, Scudder applied the new species name *obscurum* to this fossil nymphalid because it was "the most vaguely and imperfectly preserved of all our fossil butterflies, of which enough remains in sufficient preservation to indicate the family to which it belongs, and to some extent its closer affinities" (Scudder, 1889, *Eighth Annual Report, U.S. Geological Survey*, pp. 457-459, Pl. LIII, Figs. 10-13). The single fossil was collected at Florissant, apparently for Scudder's collection. It is preserved with spread wings, but only the border areas and some of the veins are preserved, along with one antenna (particularly its club), a piece of the tongue, and one of the forelegs. The atrophied nature of the foreleg is the key character placing this specimen in the Nymphalidae. This was the first fossil form found that clearly has atrophied forelegs, a character common to both sexes of modern Nymphalidae. The total expanse of the wings as preserved is 55.5 mm, and Scudder estimates the probable expanse of the intact wings as 63 mm. The body is 18 mm long. Also, the sex of this specimen can be determined: it is a male, with a part of its upper genitalia preserved at the end of the abdomen. The body is very stout, while the wings are broad and the antenna quite short with a gradually expanded club. This combination of characters is not found in any temperate North American genus today; its closest affinities seem to be with *Marpesia* and *Anaea* from Mexico and Central America.

9. *Nymphalites scudderi* Beutenmuller and Cockerell
SCUDDER'S NYMPHALID

This strange large nymphalid was described by Beutenmuller and Cockerell in 1908 (*Bulletin of the American Museum Natural History*, 24: 67, Pl.V, Fig. 6) from a single specimen collected by S. A. Rohwer at Station 14 in the Florissant fossil beds. The fossil is flattened with wings fully expanded. It is distinctive by its large size (forewing length of 39 mm, and hindwing about 36 mm long) and the sharply angled tips of the forewings. The dorsal surface is exposed, and the wings appear to be darkly colored except for a broad but rather obscure light band near the margin of both forewing and hindwing. While the markings resemble those of the Weidemeyer's Admiral *(Limenitis weidemeyerii)* that inhabits the Florissant Valley today, the shape of the forewing is different. The species name honors Samuel H. Scudder, a 19th century Harvard scientist famous for his three-volume treatise, *Butterflies of the Eastern United States and Canada, with Special Reference to New England*, and an expert on fossil insects who named eight of the twelve fossil butterflies discovered at Florissant.

10. *Apanthesis leuce* Scudder
LEUCE'S NYMPHALID

Described as a new genus and species from one specimen collected at Florissant by D. P. Long, S. H. Scudder (1889, *Eighth Annual Report, U.S. Geological Survey*, pp. 459-461, Pl. LII, Figs. 12-13) named this new species after Leuce, a nymph beloved of the god of the netherworld. Only the front wing (34.5 mm in length) is known from the fossil. It belongs to the tribe Nymphalini and perhaps is closest to *Clothilda* (=*Anelia*), a living genus in this tribe from tropical America, and *Cirrochroa*, another living genus from the East Indies in Asia, such as *C. fasciata* Felder of Sumatra. The ground color of the preserved forewing is a rich dark brown, with an intact marginal row of broad light spots and a faint series of pale cloudings between those spots and the margin. The quite unusual pattern is about as well preserved as the spotted one on the exquisite type fossil of *Prodryas persephone*.

11. *Chlorippe wilmattae* Cockerell
COCKERELL'S NYMPHALID

One of the only two Florissant butterflies assigned to living genera, this large nymphalid was originally described by T. D. A. Cockerell in 1907 (*Canadian Entomologist*, 39: 361-363, Pl. 10) from a single well-preserved type fossil collected at Station 21 (hill south of the saw mill) in July 1907 at Florissant by W. P. Cockerell, the describer's wife, and after whom the species was named. A second specimen (plesiotype) belonging to the U.S. National Museum was described by Cockerell in 1913 (*Proceedings, United States National Museum,* 44: 343, Pl. 56, Fig. 3). The venation of the hindwing is reasonably well preserved in the second fossil and is about identical to the butterflies in the living butterfly genus *Chlorippe* (=*Doxacopa*) of the American tropics. The original type fossil has a wing expanse of 64 mm (forewing length of 30 mm), and the wing markings are well preserved as a pale sepia brown in ground color, with whitish spots. These spots are arranged in a submarginal pale band, generally like modern South American species of *Chlorippe (=Doxacopa)*. However, the wing form resembles that of the living genus *Vanessa* more than that of the modern *Chlorippe*. Cockerell (1907) thought that among the fossil butterflies known from Florissant, *Chlorippe wilmattae* was most like *Lithopsyche styx* Scudder. However, the markings differ in many details and they are evidently not close relatives.

12. *Vanessa amerindica* Miller and Brown
AMERINDIAN LADY

This new Oligocene fossil butterfly was named from two specimens in 1989 by Jacqueline Y. Miller and F. Martin Brown (*Bulletin of the Allyn Museum*, No. 126: 1-9, Figs. 1-4). It is the second Florissant fossil butterfly to be placed in an existing genus.

The best preserved specimen (holotype) was one of those superb insect fossils presented to tour guides leading early groups of visitors through the area, and was given to Brown by the late Fred Bischoff. It was collected in the early 1930's from the southwestern portion of the present Monument property. The second specimen (paratype) is a superimposed left ventral forewing and hindwing, and was found on the old Stoll Ranch due north of the Monument. It was discovered in 1981 by Frederick Sanborn. The forewing length of the holotype specimen is 27 mm. All four wings and the body are preserved, with the ventral surface exposed.

The wing venation and short palpi, along with the presence of reduced forelegs, place this species in the family Nymphalidae. The pronounced tip of the forewing and a characteristic lobed shape along with the wing's venation, indicate it is in the genus *Vanessa*, and with the pattern elements visible in the forewing in these two fossils, the species is close to the living Old World Indian Lady, *Vanessa indica*, a widespread Palearctic and Indo-Malayan butterfly rather similar in pattern to the Painted Lady, *Vanessa cardui* Linnaeus, in Colorado today (as well as elsewhere in North America, northern Africa, and Europe). The name *amerindica*, therefore, refers both to the fossil's American origin and to its strong affinity with the present-day, Old World species *V. indica*. The

FIGURE 4. Fossil butterfly specimens collected in the shales of ancient Lake Florissant (not shown to scale), representing the 12 species of fossil butterflies described from Florissant. Figures are numbered as per the fossil butterfly accounts in the text. (1) *Stolopsyche libytheoides*; (2) *Oligodonta florissantensis*; (3) *Prolibythea vagabunda*; (4) *Barbarothea florissanti*; (5) *Prodryas persephone*; (6) *Lithopsyche styx*; (7) *Jupiteria charon*; (8) *Nymphalites obscurum*; (9) *Nymphalites scudderi*; (10) *Apanthesis leuce*; (11) *Chlorippe wilmattae*; (12) *Vanessa amerindica* (see also the photo facing the title page).

butterfly genus *Vanessa* today is nearly cosmopolitan, with a broad geographic range throughout the Holarctic to New Zealand, through central and southern Africa, to Central and South America, Hawaii, and east even to the small island of San Juan de Fuca east of the remote Falkland Islands (Miller and Brown 1989). Many species are migratory today. So perhaps it is not too unusual to expect a member of this highly successful, worldwide nymphalid genus to turn up 35 million years ago in central Colorado!

* * * *

The above list is complete through July 1991. Specimens of several species of fossil butterflies that have been found in the Florissant deposits are shown in the accompanying figures. It is known (F. M. Brown, personal communication) that perhaps a dozen or more other specimens were collected around the turn of the century and presented as gifts to tour guides leading groups of visitors through the area. The best-preserved specimen of *Vanessa amerindica* was a result of such collecting in the early 1930's, and it may well be that other such "souvenir" butterfly fossils will turn up in private collections in the future. In the meantime, some lucky fossil hunter may yet turn up a fossil of a new butterfly species in the Florissant beds lying outside the Monument! There were no doubt close to 100 butterfly species living in the area 35 million years ago, just as there are today. And probably they were just as abundant, before the nearby volcanic eruptions carried them to their death and subsequent decay, or occasional preservation as fossils.

THE ECOLOGY OF PRESENT-DAY BUTTERFLIES OF THE FLORISSANT REGION

DISTRIBUTION OF MODERN BUTTERFLIES

The Rocky Mountain states, especially Colorado, are rich in butterfly species. In fact, with over 250 resident, migrant, and incidental stray species recognized from within its borders, Colorado has one of the richest state butterfly faunas in North America, exceeded only by those of Texas, Arizona, and perhaps California.

This central Rocky Mountain butterfly fauna has been studied for more than a hundred years by such lepidopterists as David Bruce, William Henry Edwards, F. Martin Brown, Donald Eff, William McGuire, Bernard Rotger, Clifford D. Ferris, Ray A. Stanford, Paul R. Ehrlich, Dennis Murphy, Scott Ellis, Lowell N. Harris, James A. Scott, and many others. Yet the life cycles of even some of our common species have never been documented, or are known from just the egg or one or two larval stages described by W. H. Edwards a century ago. The ecology and evolutionary relationships of species in such complex genera as *Euphydryas* and *Oeneis* have not yet been carefully studied. With habitats ranging from warm eastern grasslands to cold alpine meadows, Colorado's 104,247 square miles present a rich variety of biotic regions for butterflies to inhabit. This book summarizes the biological and ecological information that we and our colleagues have collected to date on the butterfly species occurring in a typical montane community near Florissant, 15 miles west of Pikes Peak, in Teller County, Colorado.

Of the 97 living species recorded for the Florissant area, about 25% may be considered non-residents and strays from other localities. The remaining species display a great variety of distributional patterns in this montane area. There are no great extremes of elevation here (ranging from about 8,150' at Florissant to 9,200' on Big Blue Mountain), nor is there an unusually great number of plant communities. Yet species diversity is substantial, with an interesting mixture of common, widespread species, and rare, highly restricted species.

The distribution of butterflies in this region, as well as in other temperate-zone areas, appears to be influenced by such intrinsic physiological factors (genetically determined) as (1) degree of larval foodplant specialization, (2) the type of adult nutritional resources required (e.g., water, flower nectar, fermenting sap, or even animal feces), (3) the degree of development of wing muscles, which determines the species' ability for extended flight and migratory movements, (4) summer temperature requirements for growth and development, and degree of tolerance to winter temperature minima during hibernation, and (5) intrinsic behavioral factors, such as the lack of any dispersal tendency in the Mormon Fritillary *(Speyeria mormonia eurynome)*, which exists in one isolated colony on Big Spring Ranch but is absent from many apparently suitable habitat sites in adjacent valleys.

Various extrinsic environmental factors that also affect butterfly distribution include (1) adult and larval foodplant distributions, (2) climate, (3) elevation above sea level, (4) degree of habitat patchiness, (5) natural geographic or ecological barriers to dispersal such as mountains, rivers, and foreign habitats, and even (6) prevailing wind direction.

Availability and seasonality of water, such as in semi-permanent streams, mud banks, or seeps can influence the micro-distribution of certain lycaenid and hesperiid species that frequently sip water from damp soil.

Habitat patchiness can restrict or encourage the movement of butterflies, depending on each species' degree of habitat specialization. For example, Weidemeyer's Admiral *(Limenitis weidemeyerii)* is restricted to valleys with permanent streams, where they fly among the willows and cottonwoods. Although it is restricted to such moist areas, the river valleys serve as corridors for movement upstream and down. For other species, such as the Common Sulphur *(Colias philodice)*, dry meadows form corridors for further dispersal, while thickets and forests tend to inhibit flight. A species that can fly in a variety of habitats, such as the Dark Wood Nymph *(Cercyonis oetus)*, will reach isolated forest glades where other species are excluded by an ecological barrier such as thick Ponderosa Pine forest. Those restricted by behavioral barriers to particular ecological situations and those with very specialized foodplant requirements, such as the Thicket Hairstreak *(Mitoura spinetorum)*, which feeds on dwarf mistletoes *(Arceuthobium)*, will have much more restricted distributions than butterflies with broader tolerances and more catholic tastes.

FIGURE 5. Dry meadows form corridors of dispersal for sun-loving butterfly species to move between tongues of pine-forested ridges. Crystal Peak is in the background. Rocky soils on granitic outcrops, as in the foreground, have an impoverished flora but are rich in *Sedum lanceolatum*, visible as light-colored flowers in the center of the photograph.

HABITAT DIVERSITY AND BUTTERFLY DISTRIBUTION

The general landscape of the Pikes Peak region today is one of rolling hills. Upper slopes are covered by pine forests, and open valleys support grasses and wildflowers in abundance. In fact, about 450 species of vascular plants occur in the Florissant Fossil Beds National Monument (Edwards and Weber, 1990). Small semi-permanent streams wind through each valley. Much of our field work has been concentrated on property owned by the Colorado Outdoor Education Center, located from about four miles to nine miles southwest of the town of Florissant, Colorado, on either side of the Park County-Teller County line. This 6,000-acre tract has been declared a National Environmental Study Area by the National Park Service. Most of the terrain lies between 8,400' (2,600 meters) and 9,200' (2,800 meters) elevation. The major types of butterfly habitats that occur in this area are described below.

Wet Meadows

Along the small streams that meander through the valleys of this montane community are moist meadows that support lush vegetation during spring and summer. Groves of Quaking Aspen *(Populus tremuloides)* often border such meadows, and willows (*Salix* species) may form dense thickets at the edges of the streams. Shrubby Cinquefoil *(Pentaphylloides floribunda),* Wild Iris *(Iris missouriensis),* clovers (*Trifolium* species), louseworts (*Pedicularis* species), and Shooting Star *(Dodecatheon pulchellum)* blaze yellow, blue, red, and purple against a vivid background of green grasses and sedges. By August, many of the plants begin to senesce, and the once-green meadows turn brown, although many composites (e.g. sunflowers and asters) reach their flowering peak at this time.

FIGURE 6. A typical extensive wet meadow habitat in a flat, well-watered valley near Florissant.

Wet meadows also occur on hillsides where underground springs reach the surface and form seeps, such as near the top of Little Blue Mountain. These spring-fed meadows usually support fewer species of butterflies than those along streams in the valleys. Butterflies that frequent wet meadows include the Sonoran Skipper *(Polites sonora)*, Western Tiger Swallowtail *(Papilio rutulus)*, various sulphurs *(Colias* species*)*, the Ruddy Copper *(Lycaena rubidus)*, the Greenish Blue *(Plebejus saepiolus)*, the Field Crescentspot *(Phyciodes campestris)*, Weidemeyer's Admiral *(Limenitis weidemeyerii)*, and the Butler's Alpine *(Erebia epipsodea)*.

FIGURE 7. Riparian vegetation at the bottom of a dry canyon in the region immediately south of Cripple Creek.

FIGURE 8. Stream in aspen grove at High Trails Ranch near the base of Little Blue Mountain.

FIGURE 9. Above: A shallow pond and marsh habitat at the Top-of-the-World mountain ridge, a half-mile west of High Trails Ranch headquarters in the Florissant region.

FIGURE 10. Left: Aspens in the fall, showing a pond in front of an aspen grove at the base of Little Blue Mountain and a typical pondside habitat for the Florissant region.

FIGURE 11. Forest boundary in the background. Wet meadows that support abundant Shrubby Cinquefoil such as this one are a productive habitat for Satyrinae and Nymphalinae butterflies.

FIGURE 12. A montane dry hillside near the Pikes Peak Research Station, showing a substantial stand of *Yucca glauca* plants.

FIGURE 13. Dry meadows such as this one in the Mueller Ranch State Park occur frequently in the Florissant region. Dome Rock dominates the background.

Dry Meadows

Shrubs, herbs, and grasses also inhabit well-drained, gravelly soils of hillsides and dry valleys. Great expanses of meadow on these soils are "dry" during much of the growing season. Dry meadows are often grazed by cattle, but wildflowers such as Mariposa Lily *(Calochortus gunnisonii),* harebells *(Campanula* species), paintbrush (*Castilleja* species), scarlet gilia (*Ipomopsis* species), and loco-weeds (*Oxytropis* species) grow in profusion on these sites. Steep south-facing slopes sometimes support plants such as Brittle Cactus *(Opuntia fragilis)*, Prickly Pear *(Opuntia compressa)*, Hen and Chickens *(Echinocereus viridiflorus)*, Sulphur Flower *(Eriogonum umbellatum)*, Bitterbrush *(Purshia tridentata)*, Mountain Mahogany *(Cercocarpus montanus)*, and Spanish Bayonet *(Yucca glauca)*, which normally grow at lower elevations. Butterflies that frequent dry meadows include the Draco Skipper *(Polites draco)*, several grass skippers (*Hesperia* species), the Phoebus Parnassian *(Parnassius phoebus)*, California Hairstreak *(Satyrium californica)*, Blue Copper *(Lycaena heteronea)*, checkerspots *(Poladryas arachne* and *Euphydryas anicia)*, and arctics (*Oeneis* species).

Forests

The landscape of the Florissant area is a patchwork of meadows and forests. Ponderosa Pine *(Pinus ponderosa)* is the dominant conifer of south-facing slopes and hilltops. Mature Ponderosa Pine forest often has a park-like quality, with a grassy understory and a rather open canopy. Some butterflies, such as the Pine White *(Neophasia menapia)*, Thicket Hairstreak *(Mitoura spinetorum)*, Hoary Elfin *(Incisalia polios)*, and Western Pine Elfin *(Incisalia eryphon)*, are largely restricted to this habitat because their larvae feed either on the leaves of the pines or on plants that are associated with Ponderosa Pine.

On cooler, north-facing slopes and ravines, Douglas-fir *(Pseudotsuga menziesii)* is the dominant tree. Occasionally, small stands of Colorado Blue Spruce *(Picea pungens)* may occur in sheltered valleys. Few butterflies live in these dense, dark forests, because the flowers they need for nectar and their larval host plants rarely grow under these conditions.

FIGURE 14. Ponderosa Pines (*Pinus ponderosa*) dominate the Montane Life Zone habitat of the Florissant region today.

FIGURE 15. Above: In this photograph, an ecotone (a meeting place for diverse habitats) appears between Ponderosa Pines on a rocky slope and an aspen grove at the edge of a wet meadow, on Big Spring Ranch in the Florissant region. Butterflies reach higher diversity in such areas, with faunas of adjacent habitats being shared in the ecotone.

FIGURE 16. Gambel Oak woodland in Fremont County just south of Cripple Creek at an elevation of about 8,000'. Butterflies from this general Upper Sonoran Zone habitat move periodically north into the higher Cripple Creek montane region and southern Florissant Valley, and occur as regular strays in our area.

Quaking Aspen is the most abundant deciduous tree of montane areas in Colorado. This species commonly produces suckers from the roots and gradually forms extensive clones, which may grow to be hundreds of years old. The clones are especially noticeable in the fall when each aspen grove changes color at a slightly different time. Aspen is the larval host plant of a few butterflies, such as the Western Tiger Swallowtail *(Papilio rutulus)* and the Dreamy Dusky Wing Skipper *(Erynnis icelus)*, as well as a beautiful species of underwing moth *(Catocala)*.

Hilltops and Ravine Bottoms

The males of some butterflies frequent the tops of ridges and rocky bluffs, not to enjoy the panoramic views, but to find mates. Receptive females fly up to the hilltops where males aggregate; there they mate, then disperse to the lower slopes to lay their eggs. The Anise Swallowtail *(Papilio zelicaon)*, Mexican Cloudy Wing *(Thorybes mexicana)*, several grass skippers (*Hesperia* species), Thicket Hairstreak *(Mitoura spinetorum)*, admirals and ladies (*Vanessa* species), Arachne Checkerspot *(Poladryas arachne)*, Anicia Checkerspot *(Euphydryas anicia)*, and arctics (*Oeneis* species) are notable hilltopping butterflies of our area. The vegetation of the rocky bluffs which dot the Florissant area also differs somewhat from that of other habitats. Small numbers of Limber Pine *(Pinus flexilis)* frequently occur at the summits, and Waxflower *(Jamesia americana)*, the larval foodplant for the Spring Azure *(Celastrina ladon)*, usually grows on the sides of the bluffs. Some herbs and wildflowers are restricted to this habitat.

By contrast, the males of some other butterflies perch on vegetation in the bottoms of sunny ravines and small canyons. Receptive females fly along the ravine bottoms, encounter males, mate, then disperse to other areas to lay eggs. Snow's Skipper *(Ochlodes snowi)* displays this type of mate-locating behavior.

FIGURE 17. Rocky Bluffs arise sharply from dry meadows and pine forests. Hill-topping butterfly species ascend these slopes soon after emergence to seek mates at the top, then the females descend again and disperse to lay their eggs, while the males may "hill-top" for a number of days.

FIGURE 18. The vegetation of the south- and north-facing slopes of a mountain ridge can be quite different. This photograph illustrates the importance of differential solar insolation, and depicts the transition zone between montane forest and the pinyon-juniper zone about three miles south of Cripple Creek on the Upper Shelf Road (County Road 61).

BUTTERFLY DIVERSITY

Butterflies and moths comprise the Lepidoptera, the second largest order in the class Insecta. There are more species of insects in the world than all other organisms combined. The 4,060 species of mammals, the 9,300 species of birds, and the 250,000 species of plants pale in comparison to the more than 2,000,000 species of insects estimated to exist on this planet. Some reasons for the success of insects as a group include their small size, their ability to fly, and the fact that their lives as immatures are often completely different from their lives as adults. These combined traits allow insects to occupy a remarkable diversity of habitats, ranging from tropical forests, where they are most abundant, to arctic tundra.

About one-quarter of all kinds of insects are beetles (order Coleoptera), while the Lepidoptera (with 194,000 known living species and many more undescribed) make up approximately 10% of the world's insect fauna. The great majority of lepidopteran species are moths, many of which are quite small. Butterflies number approximately 15,000 to 20,000 species and are found on all continents except Antarctica.

Butterflies can be divided into two main groups, the skippers (superfamily Hesperioidea) and the true butterflies (superfamily Papilionoidea). The differences between these two groups will be described in the next section, but here it is sufficient to note that both butterflies and skippers, unlike most moths, are active during the day. In addition to their largely nocturnal habits, moths may be distinguished from butterflies and skippers by their generally larger and hairier bodies and by the fact that the shafts of their antennae, which may be slender or have feathery side branches, usually end in a tapered point. The antennae of butterflies and skippers are never feathery and always end in a somewhat enlarged club.

Because moths are primarily active at night, they do not rely upon wing color patterns to signal species-recognition for mating purposes. Instead, females attract males by releasing pheromones that act as external hormones. Most moths rest during the day, and the external appearance of their wing patterns has been shaped by natural selection to blend in with their surroundings. Thus, many moths have a gray or brown ground color, which is strongly mottled to enhance the camouflage effect. By contrast, the upper wing surfaces of true butterflies often display bright colors that may have been selected either for species identification and sexual recognition, or as signals of unpalatability to potential predators. Skippers are often intermediate between true butterflies and moths in many traits. Skippers have relatively broader heads and larger bodies than true butterflies, and most are dully-colored dark brown or tawny.

BUTTERFLY BIOLOGY

There are four stages in the life cycle of all Lepidoptera: **egg**, **larva**, **pupa**, and **adult**. The adult, of course, is the reproductive stage. To successfully reproduce, a male and a female of the same species must recognize each other and communicate the female's receptivity on the basis of a complicated courtship ritual, and then copulate. After a male has inseminated a female, he contributes no further to the care of their offspring. It is the female's responsibility to locate the proper foodplant on which to lay her eggs. This is a rather sophisticated process, which reveals that female butterflies can be excellent botanists. Most butterflies lay their eggs only on one or a few closely related species of plants. This situation has evolved because plants differ considerably in the chemicals and nutrients they contain. To exploit a specific plant as a larval resource, the butterfly caterpillar must have a digestive system capable of processing specific chemicals.

Most butterflies and skippers lay eggs singly on their larval foodplants, but a few lay clusters of eggs that vary in number from less than a dozen to several hundred. Eggs are attached to the foodplant with a sticky substance secreted by the female. Depending on the species, eggs are placed in a variety of positions on the plant, including the upper or lower leaf surfaces, petioles, stems, leaf and flower buds, and inflorescences.

Most eggs take from three to ten days to complete development, at which point the larva hatches. The empty eggshell is often consumed before the tiny larva begins to feed on the hostplant. Larvae from eggs laid in clusters usually feed together as a group, at least in the early instars. Larvae from eggs laid singly are usually solitary feeders. Like all insects, butterflies have an external skeleton, which must be periodically shed during the larval stage to permit incremental growth. Butterfly caterpillars usually shed their external skeletons, or skins, five times. Each growth stage is called an instar. A freshly-hatched, first-instar larva is only one or two millimeters in length, but by the time it reaches the fifth instar it may be as much as 20 to 30 millimeters in length. Thus, during its three to four-week lifespan, a caterpillar increases its length by 20 to 30-fold and its volume by about 100-fold. The caterpillar accomplishes this amazing feat by feeding almost constantly. By the time it has reached its full size, it has stored enough energy to complete the process of metamorphosis, a complex reorganization of tissues that takes place in the pupal stage.

The pupa, or chrysalis, is a rigid case-like stage that contains the larval body as it disintegrates and reorganizes into the adult butterfly. This process requires about seven to ten days to complete, after which time a portion of the upper wing pattern of the adult butterfly can be seen through the thin pupal shell. Unlike the pupae of moths, which are usually enclosed in a cocoon of silk, butterfly pupae are usually naked, or have only a small skein of silk partially enclosing the pupal case. Although the chrysalis is capable of limited movement in response to stimulation (such as of a probing predator), most are completely defenseless at this stage and are protected primarily by concealment. Many butterfly pupae are green or brown, with disruptive shapes or patterns of stripes or even metallic spots. The mature larva, when ready to pupate, crawls around until it finds a suitable place for pupation. Here it spins a small silken pad on the pupation surface and suspends itself from the pad by hooking the terminal pair of legs into the silk. Nymphalid pupae are suspended up-side-down solely from the silken pad, but most other butterflies are also attached to the substrate by a silken thread around the thorax.

When metamorphosis of the pupa is complete and the adult butterfly is ready to emerge, the pupal case splits open along special lines of weakness, and the adult pulls itself free. The butterfly suspends itself from the pupal case by holding on with its legs. It then begins to pump body fluids through the wing veins, expanding the wings. In a short time, the wings become fully hardened, and the butterfly is able to fly away. This is a delicate process that usually takes place early in the morning, before the air temperatures are so hot as to cause the wings to dry before they are fully expanded. It may take several hours before the wings are completely dry.

BUTTERFLY BEHAVIOR AND REPRODUCTION

In many species, males emerge prior to the emergence of the females by several days to a week or so. This ensures that there are plenty of males available when a female does emerge. Males spend most of their adult lives searching for receptive females. Females, on the other hand, are mated soon after emergence. Thus, the major occupation of an adult female butterfly is locating the proper foodplant and depositing her eggs. Most of the elaborate patterns and colors on the wings of butterflies are believed to serve the purpose of enabling members of the same species to recognize each other during courtship.

Butterfly courtship is an elaborate behavioral ritual almost always initiated by males. Depending on the species, males hunt for females in one of two basic ways. In the patrolling strategy, males fly constantly in search of females that may also be flying or perched on vegetation. The perching strategy involves males selecting perches on vegetation at an appropriate height, and watching for females to fly by. The next step in both behavioral strategies is for the male to give chase to the female, during which he flies near her and around her to display the patterns on his wings. In some cases, the males release an aphrodisiac pheromone, which the female detects with her antennae. If

FIGURE 19. A courting pair of checkerspots (*Euphydryas anicia*). The male is approaching the female from behind, who is resisting the mating attempt by flattening her wings out to the sides and preventing the male an easy approach. If she was still virgin and receptive to the male, she would normally raise her wings vertically over her thorax, allowing the male to approach on one side and then curve his abdomen around to copulate with her.

she is unmated or has not been mated recently, and therefore is receptive, she will land. The male lands alongside her, curling the tip of his abdomen around so that it meets the tip of her abdomen. Physical contact results in the male's claspers opening and grasping the end of the female's abdomen. Once the two butterflies are hooked together by the action of the claspers, the male inserts his aedeagus (penis). Over a period of several minutes to several hours, he transfers a spermatophore to the female.

The spermatophore is a proteinaceous sac of sperm and accessory fluids, which is produced at considerable physiological cost to the male. The combined weight of the spermatophore body and the contained fluids can represent as much as 10% of the male's body weight. Females of a few species have been shown to metabolize some of these seminal fluids to support their own metabolism or to manufacture additional eggs. In the temperate zone, where the flight season for a given species is relatively short, females emerge from the pupa with the full complement of eggs that they will lay in their lifetime. By contrast, many butterflies in the southern United States, and particularly in the tropics, emerge from the pupa with no mature eggs in their reproductive tracts. The difference appears to result from differing mortalities of the larval stages between tropical and temperate butterflies.

In the tropics, the same warm humid conditions that permit butterfly flight throughout the year, also permit the year-round supply of insect predators and parasitoids. Eggs, caterpillars, and pupae are particularly susceptible to these parasitoids and predators and there is strong selection to complete the life cycle as quickly as possible and get to the adult stage. This rapid development leaves little time for the caterpillar to store large quantities of nutrients which, through reorganization at metamorphosis, are turned into eggs for the adult female butterfly. Instead, the adult female emerges with few or no mature eggs and must acquire the necessary energy and nutrients for egg production during her lifespan. She does this by feeding on nectar, animal feces, fermenting fruit, or other substances rich in nitrogenous compounds. Many tropical butterflies have adult lifespans that range from a few weeks to several months, giving ample time to collect these nutrients and produce eggs.

In the temperate zone, however, the lifespans of butterflies are considerably shorter, from several days to a few weeks at most. This shorter lifespan, coupled in most areas and seasons with fewer nitrogen-rich adult food sources, means that it would be difficult if not impossible for adult butterflies to collect the nutrients needed for egg production. As a result, nutrients obtained by caterpillars are turned into eggs during the metamorphic process in the pupa, with the result that adult butterflies in the temperate zone feed only to support their own activities, or need not feed at all.

In the temperate zone, the number of species of insect predators and parasitoids is considerably less than in the tropics, with the result that there may be less pressure on the immature stages to complete their development in the shortest period of time. In addition, the cooler temperatures in the temperate zone, especially at night, mean that larval growth rates are much slower than in the tropics. These factors result in a relatively long larval period in temperate butterflies, which makes possible the accumulation of sufficient nutrients to finance egg production during the larval stage.

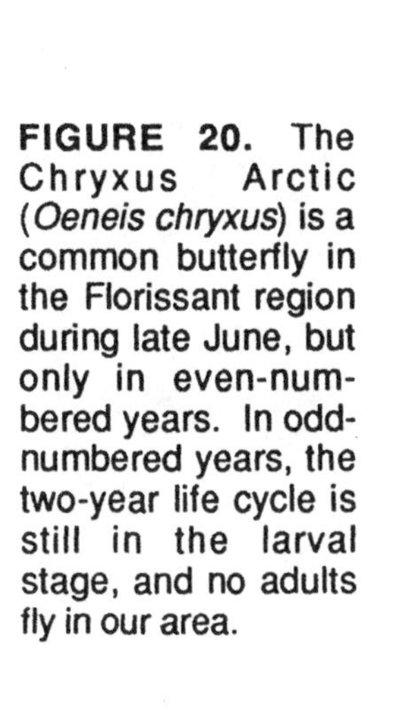

FIGURE 20. The Chryxus Arctic (*Oeneis chryxus*) is a common butterfly in the Florissant region during late June, but only in even-numbered years. In odd-numbered years, the two-year life cycle is still in the larval stage, and no adults fly in our area.

BUTTERFLY SURVIVAL

The strong seasonality of the temperate zone produces another problem for butterflies and skippers: winter cold. They must pass the winter months in a dormant stage, which may be the egg, larva, pupa, or the adult, depending on the species. Among butterflies as a group, then, all four life stages are known to enter a dormant phase that can survive extremely cold winter temperatures. It is, therefore, fascinating to speculate why different species overwinter in different stages. No doubt the reasons are complex, and probably involve historical factors, including past distributions, as well as current ecological relationships.

Even on a daily basis, butterflies in the cool montane regions are often faced with temperatures too low to initiate flight, especially in the early morning and late afternoon when temperatures are low. Butterflies and skippers, like all insects, obtain most of their body heat from their environment. Aside from muscle shivering, they do not regulate their own body heat internally. Thus, a portion of a butterfly's day is devoted to basking in the sun with wings spread open or partly open so as to gain sufficient heat energy for muscles to work in flight. Basking behavior allows butterflies to raise their body temperatures several degrees above ambient air temperature. This is a critically important behavior because there is a threshold temperature below which butterflies and skippers cannot fly. Basking, then, allows butterflies and skippers to function in spite of the climatic restrictions on behavior imposed by cool temperatures in montane environments.

Butterflies and skippers suffer greatest mortality in the immature stages, particularly the egg and larval stages. Many species of small wasps and flies specialize in parasitizing the immature stages of lepidopterans. The fact that most butterfly populations are relatively constant in size, from year to year, attests to the efficacy of these parasites. A female butterfly or skipper will lay several hundred eggs in her lifetime, and yet only a

FIGURE 21. Above: An abundance of California Hairstreaks (*Satyrium californica*) on milkweed flowers, a common sight when milkweed is in full-bloom. These flowers attract hairstreaks from the surrounding habitats in considerable numbers. **Below:** A close-up view of *Pentaphylloides floribunda*, an attractive nectaring plant for butterflies, which occurs abundantly in wet meadow habitats in the Florissant region.

small fraction of these will complete the life cycle and become adults, largely because of the parasitism and predation pressures on the immature stages. Mortality in adults is proportionately much less than that of the immature stages, largely because the flying ability of adult butterflies permits them to escape most predators. Rapid flight and good maneuverability make butterflies and skippers difficult prey for insectivorous birds or lizards.

In addition to flight agility, many species of butterflies, especially in tropical families, are protected by poisonous compounds in their wings and body tissues that render them distasteful or nauseous to predators, especially birds. Butterflies usually obtain these poisonous compounds from their larval foodplants. Larval digestive systems contain detoxifying enzymes that permit them to feed with impunity on these poisonous plant tissues and, in some cases, caterpillars are able to sequester or store these same poisonous compounds in their body, and even pass them through metamorphosis to the adult.

Distasteful butterflies generally display warning coloration in their wing patterns, usually a combination of black and yellow or orange. A good example is the Monarch butterfly, which has been shown to be toxic to some bird predators such as jays. Larvae of the Monarch feed exclusively on milkweeds (family Asclepiadaceae). Cardiac glycosides or cardenolides occur in the tissues of many milkweed species. These chemical toxins act as a heart stimulant and can poison vertebrate animals. Thus vertebrate herbivores such as deer or cattle will avoid milkweeds in a meadow. These same chemicals are sequestered in the tissues of monarch larvae and adults. Thus, the adult butterfly also is protected from its vertebrate predators by harboring poisonous compounds extracted from the larval foodplant during larval feeding.

FIGURE 22. A female Phoebus Parnassian (*Parnassius phoebus*) spreads her wings in basking behavior in the morning sunlight. This butterfly has been marked in a coded pattern on the left and right forewings, and can be identified as an individual and followed by biologists each day during its several-week lifespan, so that more can be learned about its behavior in the field. The larva is black with bright yellow spots; both the larva and the red-spotted adult may be distasteful to potential predators.

BUTTERFLY NAMES AND CLASSIFICATION

Most people who are interested in nature like to know the names of the animals and plants they are observing. In fact, for many people the challenge of identifying and labeling flowers, birds, or butterflies is the most enjoyable part of their interaction with the natural world. Others simply use names as a handle, a practical system for keeping track of their observations and a convenient way to discuss their interest with like-minded friends. Although everyone agrees that names are important, we fall short of a consensus when it comes to deciding which names to use. In the world of nature there are two kinds of names, **common names** and **scientific names.**

Many butterflies have more than one common name, especially in different parts of the country, and authors of popular guides to butterflies do not always agree on which one to use. This can lead to considerable confusion when consulting several references to learn what is known about a particular species. Scientists have long recognized this problem, however, which is why they have agreed on a system of Latin binomials to label every species known to science with a unique name combination. In this system, the first name refers to the genus and is always capitalized. The second name refers to the species and is never capitalized. Even though it is called a binomial system, you will notice in the entries below that many species have three names, not just two. The third name (which is never capitalized) refers to the subspecies, which is the distinct geographic form or race of that species in a particular area.

Subspecies of the same species rarely overlap in distribution and usually reflect the tendency of isolated populations of a species to evolve observable differences in wing pattern, foodplant use, or behavior. Some scientists consider subspecies to be the equivalent of races of the same species; others think of them as incipient species, which, given enough time in continued isolation, would evolve into separate species. At any rate, many of our western species have subspecific names because mountains, valleys, and deserts often form barriers to dispersal, fragmenting widespread species into several local populations that have had time to evolve differences during relative isolation.

Following the Latin name of the butterfly, which always is underlined or appears in italics, is the name of the person who formally described the species (for binomials) or subspecies (for trinomials). If the person's name appears in parentheses, this notation indicates that that butterfly has been assigned to a genus other than the one to which the original author assigned it, reflecting more recent (and hopefully better) scholarship.

Once a butterfly has been given a specific Latin binomial (or trinomial), it is assigned to an ascending series of higher taxonomic categories. Occasionally, very closely related species are grouped into a subgenus, such as *Chalceria* or *Epidemia* in the copper genus *Lycaena* (see page 57). Just as species are assigned to the same **genus** (or subgenus), closely related genera are grouped into **subfamilies**. Similar subfamilies comprise the **families** that serve as the major headings in the following section. The several families of true butterflies form one superfamily, the Papilionoidea, while the skippers form a separate superfamily, the Hesperioidea, and these together with a number of superfamilies of moths comprise the insect **order** Lepidoptera.

Our butterfly fauna can be divided into six families. Skippers make up the family Hesperiidae, and true butterflies are distributed among Papilionidae, Pieridae,

Lycaenidae, Riodinidae, and Nymphalidae. The last is the largest family of butterflies and has several subfamilies, which some authors treat as full families. So, though scientists agree on the binomial **system** for naming organisms, this does not mean that they always agree on what **names** should be assigned to a particular plant or animal, or how these species should be **grouped** in higher categories. Assignment of names was originally made on the basis of external morphological appearance. Later, differences in reproductive anatomy were recognized as useful in the classification of butterflies. More recently, differences and similarities in behavior, physiology, and genetics have been used to confirm or revise phylogenies (genealogical arrangements) of butterflies and other organisms. As new techniques are applied to butterfly systematics, our classification of this fascinating group of insects evolves to reflect our new understanding.

The arrangement of species in this book conforms to the sequence presented in the *Check List of the Lepidoptera North of Mexico* (Hodges 1983), although we have departed somewhat from its generic nomenclature, which is based on the controversial Miller & Brown (1981) *Catalogue/Checklist of the Butterflies of North America.* In some cases we have retained familiar generic names (indicating the generic designations from the Hodges' checklist in parentheses) and have adopted a more conservative higher taxonomy that recognizes fewer butterfly families. These departures from the published check lists reflect our view that some of the many new generic names in the Miller & Brown check list violate the need for nomenclatural stability and may be unwarranted from a taxonomic standpoint. Common names of butterflies conform to those published in *The Audubon Society Field Guide to North American Butterflies* (Pyle 1981) or in *Colorado Butterflies* (Brown, Eff and Rotger 1957). The international Lepidopterists' Society and the Xerces Society established a joint committee to compile a list of common names for all butterflies in North America. When this list is published by Smithsonian Press (as *Common Names of North American Butterflies*), it should reduce confusion by facilitating the cross-referencing of the many common names in use in various field guides and parts of the country.

Scientific and common names of plants conform to *Rocky Mountain Flora, 5th Edition* (Weber 1976), with updates by its author made during his verification of specimens collected by the Colorado Native Plant Society for the Herbarium of Florissant Fossil Beds National Monument, which is housed at Pikes Peak Research Station. A complete list of all 443 plant species identified during this project was published by Colorado Outdoor Education Center in 1990 as *Bulletin of Pikes Peak Research Station No. 2,* entitled *Plants of Florissant Fossil Beds National Monument,* by Mary F. Edwards and William A. Weber. We have given the family names of larval host plants in parentheses following the first mention of that particular host in the individual butterfly species accounts section.

THE PRESENT-DAY BUTTERFLY FAUNA OF THE FLORISSANT REGION

The modern butterflies of the Florissant region were first studied extensively by Thomas C. Emmel, starting in 1960. Following several years of field work, a total of 60 species were recorded for the area, and a list was published with distributional and ecological information (Emmel 1964). In subsequent years, many other naturalists and biologists have visited the area to participate in field activities at Colorado Outdoor Education Center, and also to contribute to the biological inventories associated with the establishment of Florissant Fossil Beds National Monument in 1969. Their activities led to the discovery of many more species in the area and additional life history information. Beginning in 1982, annual workshops on the biology of the Lepidoptera have been held at the Colorado Outdoor Education Center and Pikes Peak Research Station. Amateurs and professionals from all over North America who participated in these workshops have contributed their records to the present work.

As a result of all these efforts, we are able to offer in the following pages a current account of each of the 97 butterfly species recorded for the Florissant region. These species accounts include ecological and distributional information, annual flight periods, descriptions of immature stages, foodplant records, and additional biological notes of interest for each butterfly.

FIGURE 23. Willow Dock (*Rumex salicifolius* subspecies *triangulivalvis*), the host plant of the Ruddy Copper (*Lycaena rubidus*), grows along small streams in the Florissant area.

Superfamily HESPERIOIDEA
THE SKIPPERS AND GIANT SKIPPERS

Skippers are relatively small, stout-bodied insects that differ enough from other butterflies to warrant their placement in a separate superfamily, the Hesperioidea. Along with other, more arcane details, one easily recognizable difference between the two groups is in the shape of the antennae. Most butterfly antennae end in a distinct rounded club. In skippers, the tip of the antennal club or apiculus is bent, sometimes so much as to appear hooked. The name "skipper" derives from their fast, powerful flight, usually undertaken over short distances, and quick acceleration and brakings as when they "skip" from flower to flower. There are two families in the Hesperioidea: the Giant Skippers or Megathymidae, which occur on the plains in Colorado but not at our elevations, and the Skippers or Hesperiidae, the family which contains all our species in montane central Colorado.

Family HESPERIIDAE
THE SKIPPERS

Skippers of the family Hesperiidae are worldwide in distribution, with over 3,500 species comprising six subfamilies, four of which are represented in North America. Of the 263 North American species, most are found in the southern states. Nineteen species in two subfamilies have been recorded as residents or migrants in the Florissant area. The Pyrginae (Broad-winged Skippers) usually rest or bask with their wings open, with both forewings and hindwings on the same plane. Males of this group lack well-defined stigmas (dark patches of pheromone-disseminating scales) on the upper surface of the forewings, but many species have a costal fold (a flap on the leading edge of the wing that conceals scent-producing scales) on the forewing. Larvae feed on a wide variety of dicotyledonous plants. By contrast, the Hesperiinae (Branded Skippers) usually rest or bask in the sun with their forewings and hindwings open at different angles rather than in the same plane. They are called Branded Skippers because males of many species have clearly defined stigmas (sex patches) that appear as dark lines (brands) on the upper forewings. Another name for this group is Grass Skippers, because their larvae feed mostly on grasses and sedges (monocotyledons).

Subfamily PYRGINAE

1. ***Thorybes mexicana nevada*** **Scudder** Plate I, Fig. 1a-c
MEXICAN CLOUDY WING

The Mexican Cloudy Wing is a medium-sized skipper with dark brown wings. The forewings bear small white apical and medial spots. The undersides of the hindwings are marked with dark bands and are grayish distally. Unlike males of most *Erynnis* species, which are somewhat similar, the Mexican Cloudy Wing lacks any costal fold at the front edge of the forewings. This skipper is single brooded (passes through one generation per year) and flies in meadows from early June to early July. Males frequent hilltops and may also be found sipping water from wet soil. Both sexes visit the flowers of loco-weeds (*Oxytropis*) for nectar.

Eggs are laid singly on the leaves of legumes, especially clover (*Trifolium* species) (Fabaceae). As in most skipper butterflies, the caterpillars construct shelters by partially cutting and folding over sections of leaves or tying several leaves together and lining the interiors with silk. The mature caterpillar has a dark head and prothoracic shield. Partly grown larvae probably overwinter.

2. ***Erynnis icelus*** **(Scudder and Burgess)** — Plate I, Fig. 2-a-c
DREAMY DUSKY WING

The Dreamy Dusky Wing is a small dark skipper that lacks white spots. Instead, a chain-like band of black-bordered gray spots lies across each forewing. Male Dreamy Dusky Wings lack the costal fold found in other species of *Erynnis*. However, they do possess tibial hairpencils; these brush-like scales on the hind legs are used to disseminate pheromones during courtship. Adults can be found in low numbers in meadows from late May to early July. The flowers of Wild Iris *(Iris missouriensis)*, loco-weeds (*Oxytropis* species), and Butterweed *(Senecio eremophilus* var. *kingii)* are visited for nectar. Unlike other butterflies, some *Erynnis* species, including the Dreamy Dusky Wing, sleep with the wings held to the sides, the forewings covering the hindwings.

Quaking Aspen *(Populus tremuloides)* (Salicaceae) is the larval hostplant. Early instar larvae have black heads and green bodies with two thin yellow stripes edging the dorsal surface. The prothoracic shield is indistinct. The head is brown with a dark border in the last instar (Figure 24). Mature larvae overwinter, and then pupate in the spring without further feeding. The pupa is green with black thoracic spiracles.

3. ***Erynnis funeralis*** **(Scudder and Burgess)** — Plate I, Fig. 3a-c
FUNEREAL DUSKY WING

The Funereal Dusky Wing is a large, very darkly-colored species of *Erynnis*. The hindwings are very broad and have a contrasting white fringe. The rather long and narrow forewings have a few white spots and a pale patch at the end of the cell on the upperside. Males have costal folds. The Funereal Dusky Wing does not overwinter in montane Colorado, and rarely strays into the Florissant region. A single, very worn specimen was taken on the Big Spring Ranch on 30 July 1987.

FIGURE 24. *Erynnis icelus* (Species 2). Fourth and fifth instar larvae (Teller County, Colorado) on Quaking Aspen *(Populus tremuloides)*.

FIGURE 25. *Erynnis persius fredericki* (Species 5), last instar larva on Golden Banner *(Thermopsis divaricarpa)* (Teller County, Colorado).

4. *Erynnis afranius* (Lintner) Plate I, Fig. 4a-c
AFRANIUS DUSKY WING

The Afranius Dusky Wing is a small dark skipper with white spots. Males have costal folds and tibial hairpencils. This species is nearly identical to its close relative, *Erynnis persius*, and males of the two species are best separated by examining the structure of the genitalia. The wings of *E. afranius* tend to be more strongly patterned and lack the long grayish hairs of *E. persius*. The Afranius Dusky Wing only occasionally strays into the Florissant area from lower elevations. The foodplants of the immatures are legumes, including lupines *(Lupinus argenteus), Astragalus* species, and *Lotus purshianus* (all Fabaceae) in other Rocky Mountain localities (Scott 1986). The mature larva is pale green in ground color, dotted with tiny white points and bearing a dark mid dorsal line as well as a yellow dorsolateral line. The head is black at the edges, with tiny white dots, and has an orange face with a black central patch. Contrary to Scott (1986), there is no black collar behind the head. The larvae live in nests of folded leaves. The pupa is green. Elsewhere, the species flies usually in two broods (May and mid-summer). The hibernation stage is probably the last instar larva.

5. *Erynnis persius fredericki* H. A. Freeman Plate I, Fig. 5a-c
PERSIUS DUSKY WING

The Persius Dusky Wing flies in dry meadows in a single generation from early June to late July. *Erynnis persius* adults closely resemble *E. afranius*, but the color pattern is less distinct, and the uppersides of the forewings of males are covered with long grayish hairs. The Persius Dusky Wing is much more abundant than *E. afranius* in the Florissant area. Adults visit the flowers of loco-weeds (*Oxytropis* species) for nectar.

Both Showy Loco *(Oxytropis splendens)* and Golden Banner *(Thermopsis divaricarpa)* (Fabaceae) are fed upon by caterpillars. Mature larvae are green with two thin yellow stripes along the edge of the dorsal surface of the body (Figure 25). The prothoracic shield is indistinct. The head is mostly light grayish brown edged with darker brown. A dark brown stripe traverses the lower part of the head. The mature larvae overwinter and pupate in the spring.

6. *Pyrgus centaureae loki* Evans Plate I, Fig. 6a-c
ALPINE CHECKERED SKIPPER

A species of rare occurrence in the Florissant area is the Alpine Checkered Skipper. This butterfly is normally found in sub-alpine and alpine meadows from 9,400' (2,700 m) to 13,000' (3,900 m) in Colorado. We have one record from the National Monument taken on 12 May 1977. *Pyrgus centaureae* is one of the larger species of *Pyrgus* in our area. The wings are black with two rows of white spots. Males have costal folds and tibial hairpencils on the hind legs. Adults often visit wet soil to sip water.

The caterpillars feed on a cinquefoil, *Potentilla diversifolia* (Scott 1986), in alpine meadows, but they readily accept strawberry (*Fragaria*) (Rosaceae) leaves in the laboratory. Mature larvae have hairy, dark brown heads, light brown prothoracic shields, and brownish green bodies with faint white striations.

7. *Pyrgus xanthus* W. H. Edwards — Plate I, Fig. 7a-c
SOUTHERN CHECKERED SKIPPER

This small, dark-checkered skipper is usually uncommon. A few specimens have been taken in the National Monument from mid-May to early June. The Southern Checkered Skipper resembles *P. scriptura*, but *P. xanthus* usually has a small white spot near the base of the upper hindwings. Males lack costal folds but have tibial hairpencils on the hind legs. Adults visit the flowers of dandelion *(Taraxacum officinale)* and sip water from wet soil.

The caterpillars feed on Soft Cinquefoil *(Potentilla pulcherrima = P. gracilis* var. *pulcherrima)* (Scott 1986) in alpine meadows. The immature stages are probably very similar to those of *P. centaurea.*

8. *Pyrgus scriptura* (Boisduval) — Plate I, Fig. 8a-c
SMALL CHECKERED SKIPPER

The Small Checkered Skipper is similar to *P. xanthus* in size and coloration, but lacks a white spot at the base of the hindwings on the upper side. Adults of the spring generation, which emerge from overwintering pupae, have larger white spots and more closely resemble *P. xanthus* than individuals that emerge in the summer. Males lack costal folds but possess large tibial hairpencils on the hind legs. The hairpencils can be inserted into a special ventral pouch at the junction of the thorax and abdomen. It is thought that pheromones produced in the pouch are disseminated by the hairpencils during courtship. This butterfly lives principally along roadsides and gulches below 8,400' (2,600 m). We have one record from the first week of June from the National Monument.

The caterpillars feed on weedy Malvaceae that have pubescent leaves, such as Copper Mallow *(Sphaeralcea coccinea).* Mature larvae are green with hairy, dark brown heads and dark brown prothoracic shields. The dark brown pupa is thickly covered with white wax. Winter is passed as pupae in thin cocoons.

9. *Pyrgus communis* (Grote) — Plate I, Fig. 9a-c
COMMON CHECKERED SKIPPER

Despite its name, the Common Checkered Skipper is not very abundant in the Florissant region, perhaps because of the rarity of the larval foodplants at this altitude. This species is easily recognized by its relatively large size and the large white spots on the wings. Males have costal folds. The Common Checkered Skipper occurs in meadows from mid-June to late August. Adults sometimes sip water from wet soil. Males patrol all day for receptive females.

In our area, the caterpillars probably feed on New Mexico Mallow *(Sidalcea neomexicana)* (Malvaceae) an uncommon plant of wet meadows. At lower elevations, weedy, introduced species of cheeseweeds (*Malva*) (Malvaceae) are the principal foodplants. The greenish-white eggs are laid singly on the leaves of the host. The partially grown larvae change from green to brown as winter approaches, and they enter diapause. Feeding resumes in the spring. The mature larva is very similar to other *Pyrgus* species, green with a dark brown head and prothoracic shield. Pupae are green with numerous black spots on the dorsum.

Subfamily HESPERIINAE

10. *Oarisma garita* (Reakirt) GARITA SKIPPERLING

Plate I, 10a-c

The Garita Skipperling is one of the most common hesperiids in our area. Adults frequent meadows where they flutter slowly through the vegetation, low to the ground. The flowers of White Dutch Clover *(Trifolium repens)*, Common Wild Geranium *(Geranium caespitosum)*, and various composites are visited for nectar (Figure 26). This small brown butterfly can be found from mid-June to early August.

The larval foodplants are grasses (Poaceae), such as needle-grass (*Stipa*), blue-grass (*Poa*), and grama (*Bouteloua*), that grow abundantly in dry meadows. The caterpillar is green with white stripes (Scott 1986).

FIGURE 26. The Garita Skipperling (*Oarisma garita*), feeding on milkweed nectar. This skipper species becomes very abundant at mid-summer.

11. ***Stinga morrisoni*** **(W. H. Edwards)** Plate I, Fig. 11a-c
MORRISON'S SILVER SPIKE

This golden brown skipper resembles some *Hesperia* species, but may be recognized by the conspicuous white basal dash and chevron-like white band ("silver spike") on the underside of the hindwings. The male stigma appears as a dark stripe in the center of each forewing. Morrison's Silver Spike has a patchy distribution in the Florissant region. Adults can be found on steep south-facing slopes and ridges from late May to early June. Males typically perch on ridge tops but can also be found visiting mud puddles. Females fly on the steep slopes, laying large white eggs on grasses. The eggs do not hatch until fall. Unfortunately, the immature stages of this species are undescribed.

12. ***Hesperia uncas uncas*** **W. H. Edwards** Plate I, Fig. 12a-c
UNCAS SKIPPER

In contrast to other species of *Hesperia* that occur in the Florissant area, the Uncas Skipper is larger, darker, and has white along the veins connecting the two bands of white spots on the underside of the hindwings. Males have a long narrow stigma on the upperside of each forewing. Adults fly in dry meadows in low abundance during July and August.

Females lay white eggs singly on bunchgrasses such as grama (*Bouteloua*) and needle grass (*Stipa*). The larvae live in silk tubes at the base of the hostplant. Mature caterpillars are brown with black heads (Figure 27). The pupa is greenish-white with some dark markings on the dorsum.

13. ***Hesperia comma ochracea*** **Lindsey** Plate I, Fig. 13a-c
COMMON BRANDED SKIPPER

This medium-sized skipper flies in dry meadows in August, several weeks later than the similar *Hesperia nevada*. The underside of the hindwings is yellowish in male *Hesperia comma ochracea* and greenish in the females. Compared to other forms of this species, the outer band of white spots tends to be reduced in size in this subspecies, especially in males. A long thin stigma is present on each forewing of the males. Both sexes visit composites such as Black-eyed Susan *(Rudbeckia hirta)* for nectar, but males are more often encountered perching on hilltops.

The white eggs are laid singly on grasses. The larval body is brown, and the head is mostly black but brownish toward the sides, and has two short parallel white lines at the top.

14. *Hesperia nevada* (Scudder) Plate I, Fig. 14a-c
NEVADA SKIPPER

The Nevada Skipper is a common inhabitant of montane grasslands in western North America. This species is smaller and emerges earlier in the season (mid-June to late July) than other *Hesperia* species in our area. The underside of the hindwing is dark green with large white spots. Males have a conspicuous dark stigma on the upperside of the forewings. Adults often visit composites, such as thistles (*Cirsium*), for nectar, but males can also be found perching on hilltops.

The larval foodplants are bunchgrasses. The eggs are white and about 1.4 mm in diameter (MacNeill 1975). The caterpillars are dark brown with black heads and prothoracic shields. There are several pale tan blotches on the larval face. The partly grown larvae diapause through the fall and winter.

15. *Polites draco* (W. H. Edwards) Plate I, Fig. 15a-c
DRACO SKIPPER

The Draco Skipper is one of the most common skippers in the Florissant region. Adults fly in dry meadows from mid-June to mid-July. This species is easily distinguished by the large and irregular, yellowish spots on the underside of the hindwings. Males bear a dark stigma on each forewing.

Various blue-grasses (*Poa*) are fed upon by the larvae, which are dark brown with black heads and prothoracic shields. The body is covered with small dark spots, and the heart is visible as a dark line through the transparent dorsum. As in many *Polites* species, the larvae also have some conspicuous dark markings on the suranal plate (Figure 28). The pupae are black with dark brown abdomens and are thinly coated with white wax.

16. *Polites themistocles* (Latreille) Plate I, Fig. 16a-c
TAWNY-EDGED SKIPPER

This small brown skipper is plainly marked on the ventral surfaces, but both sexes bear a tawny patch along the upper costal margin of the forewings. Males have short, dark stigmas on the forewings. The Tawny-edged Skipper flies in wet meadows in low abundance from mid-June to mid-July. The flowers of loco-weeds (*Oxytropis*) and White Dutch Clover *(Trifolium repens)* are visited by adults for nectar.

The pinkish eggs are laid singly on grasses. The body of the last instar larva is brown with small, dark spots. Vague dark lines run down each side, and the heart is distinctly visible as a dark dorsal line (Figure 29). The head and prothoracic shield are black, and the suranal plate is whitish or light brown with dark markings. In the Florissant area, this species probably overwinters as partially grown larvae. The pupae are greenish with light brown abdomens.

FIGURE 27. The last instar larva of *Hesperia uncas uncas* (Species 12), from El Paso County, Colorado reared on *Poa pratensis* (Kentucky Bluegrass).

FIGURE 28. Left: *Polites draco* (Species 15), last instar larva (Teller County, Colorado) reared on *Poa pratensis* (Kentucky Bluegrass).

FIGURE 29. Below: The last instar larva of *Polites themistocles* (Species 16), from Teller County, Colorado, on *Poa pratensis*.

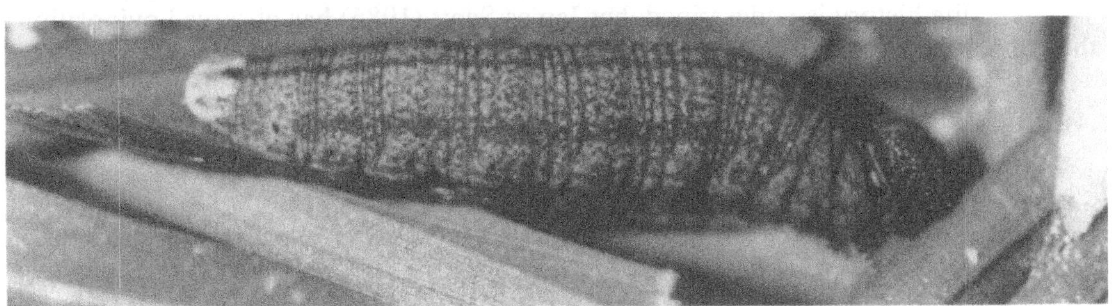

17. ***Polites sonora utahensis* (Skinner)** — Plate I, Fig. 17a-c
SONORAN SKIPPER

The Sonoran Skipper occurs in wet meadows from early July to mid-August. Males bear thick dark stigmas. Both sexes are greenish ventrally, with a white basal dash and an outer chain of small white spots. Adults avidly visit composite flowers such as asters (*Aster*) and thistles (*Cirsium*) for nectar.

The white eggs are laid on grasses. Last instar larvae are brown with vague dark spots, the heart is visible as a dark line on the dorsum. The suranal plate is colored similarly to the rest of the body, but the head and prothoracic shield are black. Pupae are black with dark brown abdomens and are covered with a thin layer of white wax.

18. ***Atalopedes campestris huron* (W. H. Edwards)** — Plate I, Fig. 18 a-c
SACHEM OR FIELD SKIPPER

The Sachem or Field Skipper may be fairly abundant some years at lower elevations in Colorado. In the Florissant region, it is usually found in low abundance from late July to early August. Males are easily recognized by the large oval stigma on the forewings. Females resemble *Hesperia* species, but may be distinguished by the squarish, translucent spot in the middle of the forewings.

The caterpillars feed on grasses, and the Field Skipper frequently breeds in suburban lawns at lower elevations. The mature larva is brown with a black head and prothoracic shield. The heart is visible as a dark line, and the body is covered with small dark spots. The pupa is mostly dark brown, but the light brown abdomen is covered with dark brown spots.

19. ***Ochlodes snowi* (W. H. Edwards)** — Plate I, Fig. 19a-c
SNOW'S SKIPPER

Snow's Skipper is a medium-sized, dark brown hesperiid that flies from mid-July to early August. The ventral wing surfaces are reddish with a row of small yellow spots. Males have a dark stigma on the forewings. In our area, Snow's Skipper has been found sparingly in wet meadows, but males typically perch in the bottom of ravines to await females. This species has also been collected at ultra-violet lights at night on several occasions (Emmel 1964).

The life history is undescribed, but James Scott (1986) has observed oviposition on the grass, Pine Dropseed *(Blepharoneuron tricholepis).*

20. ***Euphyes ruricola metacomet* (Harris)** Plate I, Fig. 20a-c
DUN SKIPPER

The Dun Skipper is a remarkably plain, dark brown skipper. Males are nearly unmarked except for the black stigma on the forewings. Females have a few light spots on the forewings and vague light spots on the underside of the hindwings. Both sexes have many yellow scales on the head and palpi. Although common at lower elevations, the Dun Skipper is rather rare in the Florissant region. Adults can be found perching on vegetation along streams and visiting flowers.

Females are known to oviposit on sedges such as Sun Sedge *(Carex heliophila)* (Cyperaceae) in other areas (Scott 1986). The eggs are greenish white with a red swirl. The caterpillar is green with a net-like pattern of white. Its light-colored head is marked with brown stripes and an oval black spot, giving it a cyclopian appearance. The prothoracic shield is narrowly banded with black, and the suranal plate is often edged with black and frequently bears a black crescent. The larvae live in the center of the hostplant rosette and make shelters by tying the edges of the leaves together and lining the tube with silk. This species overwinters as partially grown larvae.

Superfamily PAPILIONOIDEA
THE TRUE BUTTERFLIES

The superfamily Papilionoidea includes all butterflies other than skippers and numbers well over 12,000 species worldwide. Papilionoid butterflies as a group are more colorful than skippers and usually have larger wings and thinner bodies. A distinguishing character between the two groups is the shape of the clubbed antenna, which, as mentioned previously, is straight in butterflies and bent in skippers. Of the 416 species of true butterflies that occur in North America, 77 have been recorded in our area.

Family PAPILIONIDAE
THE PARNASSIANS AND SWALLOWTAILS

The family Papilionidae numbers only 566 species worldwide (Shields 1989b), but what it lacks in number of species, it more than makes up for in individual size. Papilionids are generally quite large, and the family includes the largest butterflies in the world. Two of the three Papilionidae subfamilies occur in our region, the Parnassiinae (parnassians), with one species, and the Papilioninae (swallowtails), with five species. Adults of both sexes are characterized by having six walking legs of about the same size, a trait shared with the Pieridae. The larvae lack spines or hard horns, but are equipped with an osmeterium, a forked fleshy organ normally concealed just behind the head. The osmeterium is everted when the larva is disturbed, and emits a foul smell. Some larvae have large eyespots that confer a resemblance to a snake's head, a deception enhanced by the forked-tongue appearance of the everted osmeterium. The larvae feed on many families of dicotyledonous plants, although considerable foodplant specialization occurs at the generic level. Papilionids are strong and fast fliers, but migration to new areas is uncommon. All species visit flowers for nectar and bask in the sun with wings spread open. Members of the Papilioninae (swallowtails) are unique for continuing to flutter their wings when visiting flowers. The name "swallowtail" comes from the graceful tail-like extensions on the hindwings of many species.

Subfamily PARNASSIINAE

21. ***Parnassius phoebus sayii* W. H. Edwards** — Plate II, Fig. 21a-b
PHOEBUS PARNASSIAN

The Phoebus Parnassian is widely distributed in the montane and alpine regions of Colorado. In the Florissant area, this butterfly occurs most commonly in dry meadows that support an abundance of the larval foodplant, Stone Crop *(Sedum lanceolatum)* (Crassulaceae). Males begin to emerge in early June, about a week earlier than the females. Females tend to be much less abundant than males and are most common late in the flight season. During the day, males patrol constantly for virgin females with which to mate. The delicate courtship rituals of other butterflies are unknown to these aggressive males, who attempt to wrestle any female to the ground and copulate with her. After a spermatophore (sperm sac) has been implanted, the male secretes a liquid that quickly hardens to form a strong, light-weight, grayish plug, the sphragis, which functions like a medieval chastity belt to prevent further matings. Thus, females will mate only once, unless another suitor is successful in disrupting a mating pair or in tearing off an intact sphragis with his claspers (the pair of clasping organs at the tip of the male's abdomen). Both sexes have an interesting sweetly acrid odor (to the human nose), and they probably use pheromones for communication to some extent. The adults feed infrequently at the small yellow flowers of Stone Crop and butterweed (*Senecio*) and live for only one to two weeks (Figure 30).

Females deposit large white eggs singly at the base of grasses and other vegetation in the vicinity of the larval host, Stone Crop. The eggs undergo diapause until cool fall temperatures cue the larvae to emerge. The tiny caterpillars then seek out Stone Crop plants, which are often some distance away, and begin to feed. When they are not feeding, the larvae conceal themselves among rocks and in duff on the ground, making them difficult to find. The partially grown larvae overwinter and finish their growth the following spring. Mature caterpillars are velvety black with a row of bright yellow spots along each side, and are sparsely covered with black hairs. Like all North American swallowtails, the Phoebus Parnassian larva bears a forked, eversible, osmeterium organ behind the head, although it is rarely used. The pupa is formed in debris near the host plant, and is enclosed in a lightly-woven silk cocoon.

FIGURE 30. A male (left) and a female (above) *Parnassius phoebus*, basking and perched on a Stone Crop (*Sedum*) flower head and a butterweed *(Senecio)* flower, respectively.

Subfamily PAPILIONINAE

22. ***Papilio polyxenes asterius*** **Stoll** — Plate II, Fig. 22a-b
EASTERN BLACK SWALLOWTAIL

The Eastern Black Swallowtail reaches its upper elevational limit in the Florissant region. Adults are usually uncommon and are generally found as single individuals. Despite the scarcity of adults, however, the immature stages may be fairly abundant. The eggs and larvae of this species can be found by searching through inflorescences and foliage of Giant Angelica *(Angelica ampla)* and Cow Parsnip *(Heracleum sphondylium* ssp. *montanum)* plants in July. These large members of the carrot family (Apiaceae) frequently grow in wet meadows and along valley streams.

Early instar larvae of all *Papilio* species in our area are brown and white, and resemble bird droppings. Young larvae of the Eastern Black Swallowtail prefer to feed on young leaves and flowers of Giant Angelica and Cow Parsnip, but later switch to feeding on the seeds. During the last two instars, the larvae are green with yellow-spotted black bands, and the head is striped with black. When ready to pupate, swallowtail caterpillars attach themselves to twigs, stems, or rocks in a characteristic fashion. The caterpillar first spins a thin layer of silk over the pupation site. Then a small silk pad is spun near the bottom of this area. Finally, a silk thread is spun around the thorax to support the upper body. At the appropriate time, the larval skin is shed, and tiny hooks at the tip of the abdomen anchor the pupa to the silk pad. Eastern Black Swallowtail pupae occur in two color phases, brown with darker striations and green with yellow markings. David A. West and Wade Hazel (1979) discovered that pupal color is correlated with background color and texture so that green pupae predominate if the caterpillars pupate on live vegetation, while brown pupae are formed on dried stems and leaves. This species overwinters in the pupal stage.

23. ***Papilio zelicaon nitra*** **W. H. Edwards** — Plate II, Fig. 23a-c
ANISE SWALLOWTAIL

The Anise Swallowtail is relatively uncommon in our area, although it may be locally abundant elsewhere in Colorado. Most adults from Florissant are of the typical yellow form, but on rare occasions individuals of the black form, which resembles the Eastern Black Swallowtail, have been seen. The males are most often found at hilltops.

The immature stages of the Anise Swallowtail are very similar to those of the Eastern Black Swallowtail, a closely related species. The larval foodplant in the Florissant region is probably Yellow Mountain Parsley *(Pseudocymopterus montanus)* (Apiaceae), which grows in dry meadows and aspen groves.

24. ***Papilio rutulus rutulus*** **Lucas** Plate II, Fig. 24a-b
WESTERN TIGER SWALLOWTAIL

The Western Tiger Swallowtail is the most abundant swallowtail in the Florissant area. It is usually found in wet meadows, where it flies with a gliding flight among Quaking Aspens *(Populus tremuloides)* and willows (*Salix*) (Salicaceae), the larval foodplants. Adults emerge from overwintering pupae in early June and fly in a single brood until late July. Worn individuals sometimes fade considerably and may be mistaken for the Pale Swallowtail in flight.

Females lay large greenish eggs singly on the undersides of the host leaves. Mature larvae are green, with a pair of yellow eyespots dotted with black and blue pupils on the enlarged second thoracic segment (Figure 31). A narrow yellow band, followed by a narrow black band, borders the posterior of the last thoracic segment. The enlarged thorax and false eyes make the caterpillar resemble a small snake. Larvae which are ready to pupate change from green to dark brown, an adaptation which helps conceal them as they wander about, seeking pupation sites away from the host plants. The pupa is light brown, with a dark brown stripe down each side (Figure 32).

25. ***Papilio multicaudatus*** **Kirby** Plate II, Fig. 25a-b
TWO-TAILED SWALLOWTAIL

The Two-tailed Swallowtail is the largest butterfly in our area. This species may be separated from the Western Tiger Swallowtail by its much larger size, narrower black stripes and the well-developed multiple tails on the hindwings. The Two-tailed Swallowtail is found primarily in foothill canyons, but it may be fairly abundant in the Florissant area during favorable years.

The larval foodplant, wild cherry (*Prunus*) (Rosaceae), is relatively uncommon at high elevations. The immature stages are quite similar to those of the closely related *P. rutulus*. Adults begin to emerge from overwintering pupae in early July.

26. ***Papilio eurymedon*** **Lucas** Plate II, Fig. a-b
PALE SWALLOWTAIL

The Pale Swallowtail generally resembles *P. rutulus*, but the ground color is white and the stripes are broader. This species is usually found in moist canyons at lower elevations. In the Florissant region, the Pale Swallowtail is relatively rare, perhaps because the larval foodplant, Buckbrush *(Ceanothus fendleri)* (Rhamnaceae) does not occur at these high elevations. The early stages are similar to those of *P. rutulus*, a closely related species.

FIGURE 31. The last instar larva of *Papilio rutulus rutulus* (Species 24), from Teller County, Colorado, on Quaking Aspen.

FIGURE 32. The pupa of *Papilio rutulus rutulus* (Species 24), found in Teller County, Colorado, on Quaking Aspen.

FIGURE 33. The Pipevine Swallowtail, *Battus philenor philenor* (Linnaeus): adult male (left) and female (right), dorsal and ventral surface. This species has been captured in the Florissant area on two occasions to date.

27. *Battus philenor philenor* (Linnaeus) PIPEVINE SWALLOWTAIL

Text Fig. 33

The Pipevine Swallowtail is a member of the Troidini tribe of swallowtails, a predominantly tropical worldwide group whose larvae feed on pipevines in the genus *Aristolochia* (Aristolochiaceae). *Battus philenor* occurs throughout the eastern U.S., particularly in the south, and then west along the Gulf Coast and Texas to Arizona and parts of California, and further south into Veracruz, Mexico. It is an uncommon species in Colorado. Occasionally, small breeding colonies become established in the plains area. The first specimen taken in the Florissant area was found on June 29, 1989, by Dr. James L. Nation. A second one was taken at Little Blue Mountain in early July 1990 by James Scharmberg.

No native hostplants occur in the Pikes Peak region, but the larvae sometimes accept ornamental pipevines growing in gardens. The caterpillars are brownish-red with long, fleshy tubercles in rows along the body. The pupa is brown to green, with yellow splotches on the abdomen. The butterfly is multiple brooded elsewhere.

Rocky Mountain region, Hovanitz (1950) found that most populations of the Orange Sulphur have about 20% albinic females. Around Florissant, the frequency of white females is somewhat less, at about 16% (Emmel 1964). Adults commonly visit the flowers of loco-weeds (*Oxytropis*), asters (*Aster*), butterweeds (*Senecio*), thistles (*Cirsium*), and sunflowers (*Helianthus*) for nectar.

The immature stages and the biology of the Orange Sulphur are very similar to those of the Common Sulphur *(C. philodice)*, a closely related species. The mature larva is green with a broad white and narrow reddish stripe on each side, and a white subdorsal line (Figure 35). The pupa is green with a short pointed head.

36. *Colias alexandra alexandra* W. H. Edwards — Plate III, Fig. 36a-d
QUEEN ALEXANDRA'S SULPHUR

Queen Alexandra's Sulphur is the largest species of *Colias* found in our area. Males are lemon yellow with black borders above and grayish green below. Females have very reduced, diffuse borders and may be either yellow or, rarely, white. The frequency of albinic females around Florissant is much lower (less than 1%) than the 6% Hovanitz (1950) reported for western Colorado populations. Adults occur abundantly in meadows from late June to early August.

Legumes (Fabaceae) such as milk vetch (*Astragalus*), loco-weed (*Oxytropis*), Golden Banner (*Thermopsis*), clover (*Trifolium*), and vetch (*Vicia*) are eaten by the immature stages. The mature larvae are green, with white and orange lateral stripes. The body is covered with short hairs and small black points.

37. *Eurema nicippe* (Cramer) — Plate III, Fig. 37a-c
SLEEPY ORANGE

The Sleepy Orange is readily identified by its deep orange color and wide black margins. The larval host plants, several species of senna *(Cassia)* (Fabaceae), do not occur in our area, preventing permanent establishment of this butterfly. Worn adults of this species, which is sometimes called the Rambling Orange, occasionally move into the Florissant region during June and July.

38. *Eurema mexicana* (Boisduval) — Plate III, Fig. 38a-c
MEXICAN YELLOW

The wide and greatly uneven black borders on the forewings, and short tail on the hindwings, readily distinguish the Mexican Yellow from other yellow butterflies.

COLOR PLATES

The following nine pages of color plates depict all the known living butterflies of the Florissant area in central Colorado. The figures and corresponding figure legends are numbered as in the species accounts in the text. Each legend includes the name of the butterfly, the author who named it, the surface shown (dorsal or ventral), the sex, and the locality where the figured specimen was collected. Because perfect specimens were not always available from the Florissant area for illustration, some geographically widespread species are illustrated by specimens from other locations.

COLOR PLATE I

1a. *Thorybes mexicana nevada* Scudder, dorsal male (Quick Ranch, SW of Florissant, Teller Co., Colorado).
1b. *Thorybes mexicana nevada* Scudder, dorsal female (Pikes Peak Research Station, Teller Co., Colorado).
1c. *Thorybes mexicana nevada* Scudder, ventral male (Top of the World, SW of Florissant, Teller Co., Colorado).
2a. *Erynnis icelus* (Scudder and Burgess), dorsal male (Pikes Peak Research Station, Teller Co., Colorado).
2b. *Erynnis icelus* (Scudder and Burgess), dorsal female (Pikes Peak Research Station, Teller Co., Colorado).
2c. *Erynnis icelus* (Scudder and Burgess), ventral male (Quick Ranch, SW of Florissant, Teller Co., Colorado).
3a. *Erynnis funeralis* (Scudder and Burgess), dorsal male (The Teepees, Jefferson Co., Colorado).
3b. *Erynnis funeralis* (Scudder and Burgess), dorsal female (Kleberg Co., Texas).
3c. *Erynnis funeralis* (Scudder and Burgess), ventral male (Kleberg Co., Texas).
4a. *Erynnis afranius* (Lintner), dorsal male (Mother Cabrini Shrine, Rt. 40 W of Denver, Colorado).
4b. *Erynnis afranius* (Lintner), dorsal female (Mother Cabrini Shrine, Rt. 40 W of Denver, Colorado).
4c. *Erynnis afranius* (Lintner), ventral male (Mother Cabrini Shrine, Rt. 40 W of Denver, Colorado).
5a. *Erynnis persius fredericki* H. A. Freeman, dorsal male (Quick Ranch, SW of Florissant, Teller Co., Colorado).
5b. *Erynnis persius fredericki* H. A. Freeman, dorsal female (Pikes Peak Research Station, Teller Co., Colorado).
5c. *Erynnis persius fredericki* H. A. Freeman, ventral male (4 mi. S of Cripple Creek on C.R. 61, Teller Co., Colorado).
6a. *Pyrgus centaureae loki* Evans, dorsal male (Horseshoe Mtn., W of Fairplay, Park Co., Colorado).
6b. *Pyrgus centaureae loki* Evans, dorsal female (Horseshoe Mtn., W of Fairplay, Park Co., Colorado).
6c. *Pyrgus centaureae loki* Evans, ventral male (Horseshoe Mtn., W of Fairplay, Park Co., Colorado).
7a. *Pyrgus xanthus* W. H. Edwards, dorsal male (Nr. Fairplay, Park Co., Colorado).
7b. *Pyrgus xanthus* W. H. Edwards, dorsal female (Elk Creek, Jefferson Co., Colorado).
7c. *Pyrgus xanthus* W. H. Edwards, ventral male (Jenney Springs, New Mexico).
8a. *Pyrgus scriptura* (Boisduval), dorsal male, spring form (Foster Ranch, El Paso Co., Colorado).
8b. *Pyrgus scriptura* (Boisduval), dorsal female, summer form (Foster Ranch, El Paso Co., Colorado).
8c. *Pyrgus scriptura* (Boisduval), ventral male (Foster Ranch, El Paso Co., Colorado).
9a. *Pyrgus communis* (Grote), dorsal male (Quanah at Georgia-Pacific Plant, Hardeman Co., Texas).
9b. *Pyrgus communis* (Grote), dorsal female (Rock Creek Canyon, S of Colorado Springs, El Paso Co., Colorado).
9c. *Pyrgus communis* (Grote), ventral male (Quanah at Georgia-Pacific Plant, Hardeman Co., Texas).
10a. *Oarisma garita* (Reakirt), dorsal male (Wild Goat Mtn., SW of Florissant, Park Co., Colorado).
10b. *Oarisma garita* (Reakirt), dorsal female (Quick Ranch, SW of Florissant, Teller Co., Colorado).
10c. *Oarisma garita* (Reakirt), ventral male, (Pikes Peak Research Station, Teller Co., Colorado).
11a. *Stinga morrisoni* (W. H. Edwards), dorsal male (Phantom Canyon Rd. at Mile Marker 18 1/4, Fremont Co., Colorado).
11b. *Stinga morrisoni* (W. H. Edwards), dorsal female (Phantom Canyon Rd. at Mile Marker 18 1/4, Fremont Co., Colorado).

11c. *Stinga morrisoni* (W. H. Edwards), ventral male (Phantom Canyon Rd. at Mile Marker 18 1/4, Fremont Co., Colorado).
12a. *Hesperia uncas uncas* W. H. Edwards, dorsal male (Dome Rock, 7 mi. S of Florissant, Teller Co., Colorado).
12b. *Hesperia uncas uncas* W. H. Edwards, dorsal female (Dome Rock, 7 mi. S of Florissant, Teller Co., Colorado).
12c. *Hesperia uncas uncas* W. H. Edwards, ventral female (4 mi. S of Cripple Creek on C.R. 61, Teller Co., Colorado).
13a. *Hesperia comma ochracea* Lindsey, dorsal male (Pikes Peak Research Station, Teller Co., Colorado).
13b. *Hesperia comma ochracea* Lindsey, dorsal female (Pikes Peak Research Station, Teller Co., Colorado).
13c. *Hesperia comma ochracea* Lindsey, ventral male (Pikes Peak Research Station, Teller Co., Colorado).
14a. *Hesperia nevada* (Scudder), dorsal male (Wild Goat Mtn., SW of Florissant, Park Co., Colorado).
14b. *Hesperia nevada* (Scudder), dorsal female (Quick Ranch, SW of Florissant, Teller Co., Colorado).
14c. *Hesperia nevada* (Scudder), ventral male (Quick Ranch, SW of Florissant, Teller Co., Colorado).
15a. *Polites draco* (W. H. Edwards), dorsal male (Quick Ranch, SW of Florissant, Teller Co., Colorado).
15b. *Polites draco* (W. H. Edwards), dorsal female (Quick Ranch, SW of Florissant, Teller Co., Colorado).
15c. *Polites draco* (W. H. Edwards), ventral male (Quick Ranch, SW of Florissant, Teller Co., Colorado).
16a. *Polites themistocles* (Latreille), dorsal male (Phantom Canyon Rd. at Mile Marker 18 1/4, Fremont Co., Colorado).
16b. *Polites themistocles* (Latreille), dorsal female (Holiday, Johnson Co., Kansas).
16c. *Polites themistocles* (Latreille), ventral male (Pikes Peak Research Station, Teller Co., Colorado).
17a. *Polites sonora utahensis* (Skinner), dorsal male (Pikes Peak Research Station, Teller Co., Colorado).
17b. *Polites sonora utahensis* (Skinner), dorsal female (Dome Rock, 7 mi. S of Florissant, Teller Co., Colorado).
17c. *Polites sonora utahensis* (Skinner), ventral male (Pikes Peak Research Station, Teller Co., Colorado).
18a. *Atalopedes campestris huron* (W. H. Edwards), dorsal male (9 mi. W of Rd. 164 on US 287, Childress Co., Texas).
18b. *Atalopedes campestris huron* (W. H. Edwards), dorsal female (Mitchelville, Prince George's Co., Maryland).
18c. *Atalopedes campestris huron* (W. H. Edwards), ventral male (Collier-Seminole St. Park, Collier Co., Florida).
19a. *Ochlodes snowi* (W. H. Edwards), dorsal male (4 mi. S of Cripple Creek on C.R. 61, Teller Co., Colorado).
19b. *Ochlodes snowi* (W. H. Edwards), dorsal female (Rock Creek, El Paso Co., Colorado).
19c. *Ochlodes snowi* (W. H. Edwards), ventral male (4 mi. S of Cripple Creek on C.R. 61, Teller Co., Colorado).
20a. *Euphyes ruricola metacomet* (Harris), dorsal male (Phantom Canyon Rd. at Mile Marker 11.5, Fremont Co., Colorado).
20b. *Euphyes ruricola metacomet* (Harris), dorsal female (Rock Creek Canyon S of Colorado Springs, El Paso Co., Colorado).
20c. *Euphyes ruricola metacomet* (Harris), ventral male (Broadmoor, 1/2 mi. N of Cheyenne Mtn. Zoo, El Paso Co., Colorado).

Left: The mature larva (fifth instar) of the Anicia Checkerspot, *Euphydryas anicia capella*, feeding on Indian Paintbrush (*Castilleja*).

Euphydryas anicia surface while feeding on lorissant.

ɛrspot showing her dorsal

COLOR PLATE I

COLOR PLATE II

21a. *Parnassius phoebus sayii* W. H. Edwards, dorsal male (Quick Ranch, SW of Florissant, Teller Co., Colorado).
21b. *Parnassius phoebus sayii* W. H. Edwards, dorsal female (Big Springs Ranch, Florissant, Teller Co., Colorado).
22a. *Papilio polyxenes asterius* Stoll, dorsal male (Dome Rock, 7 mi. SW of Florissant, Teller Co., Colorado).
22b. *Papilio polyxenes asterius* Stoll, dorsal female (Dome Rock, 7 mi. SW of Florissant, Teller Co., Colorado).
23a. *Papilio zelicaon nitra* W. H. Edwards, dorsal male (Magnolia Rd., Boulder Co., Colorado).
23b. *Papilio zelicaon nitra* W. H. Edwards, dorsal female (Genessee Park, Jefferson Co., Colorado).
23c. *Papilio zelicaon nitra* W. H. Edwards, dorsal male, dark form (Mt. Zion, Lookout Mtn., Jefferson Co., Colorado).
23d. *Papilio zelicaon nitra* W. H. Edwards, dorsal female, dark form (Mt. Zion, Lookout Mtn., Jefferson Co., Colorado).
24a. *Papilio rutulus rutulus* Lucas, dorsal male (Phantom Canyon Rd. at Mile Marker 18 1/4, Fremont Co., Colorado).
24b. *Papilio rutulus rutulus* Lucas, dorsal female (Quick Ranch, SW. of Florissant, Teller Co., Colorado).
25a. *Papilio multicaudatus* Kirby, dorsal male (Canon City at Red Canon Park, Fremont Co., Colorado).
25b. *Papilio multicaudatus* Kirby, dorsal female (Denver, Colorado).
26a. *Papilio eurymedon* Lucas, dorsal male (Phantom Canyon Rd. at Mile Marker 8, Fremont Co., Colorado).
26b. *Papilio eurymedon* Lucas, dorsal female (100' N. of Bears Den, Evergreen, Colorado).
96a. *Danaus plexippus plexippus* (Linnaeus), dorsal male (Elliott Key, Dade Co., Florida).
96b. *Danaus plexippus plexippus* (Linnaeus), dorsal female (Markham Co. Park, Broward Co., Florida).

Above: Rolling mountains, pine-forested ridges, and grassy valleys of the Florissant region.

Left: The Florissant region during midsummer, a carpet of rolling fields of flowers below pine- and aspen-covered ridges.

Above: Lepidopterists discuss their catch near Dome Rock in the Florissant Valley, just south of Florissant Fossil Beds National Monument.

Left: Collecting butterflies on a mountain slope south of Cripple Creek, Colorado, a thrill that can be yours for the asking virtually anywhere in the Florissant region.

COLOR PLATE II

COLOR PLATE III

28a. *Neophasia menapia* (C. and R. Felder), dorsal male (7.5 mi. S of Deckers, Hwy 67, Douglas Co., Colorado).
28b. *Neophasia menapia* (C. and R. Felder), dorsal female (Boulder, Flagstaff Mtn., Colorado).
28c. *Neophasia menapia* (C. and R. Felder), ventral male (Pine, Colorado).
29a. *Pontia sisymbrii elivata* (Barnes and Benjamin), dorsal male (Golden, Mt. Zion, Jefferson Co., Colorado).
29b. *Pontia sisymbrii elivata* (Barnes and Benjamin), dorsal female (Golden, Mt. Zion, Jefferson Co., Colorado).
29c. *Pontia sisymbrii elivata* (Barnes and Benjamin), ventral male (Golden, Mt. Zion, Jefferson Co., Colorado).
30a. *Pontia protodice* (Boisduval and Leconte), dorsal male (Pikes Peak Research Station, Teller Co., Colorado).
30b. *Pontia protodice* (Boisduval and Leconte), dorsal female (Horseshoe Mtn. W of Fairplay, Park Co., Colorado).
30c. *Pontia protodice* (Boisduval and Leconte), ventral male (Pine Island Ridge, Davie, Broward Co., Florida).
31a. *Pontia occidentalis occidentalis* (Reakirt), dorsal male (Horseshoe Mtn. W of Fairplay,' Park Co., Colorado).
31b. *Pontia occidentalis occidentalis* (Reakirt), dorsal female (Horseshoe Mtn. W of Fairplay, Park Co., Colorado).
31c. *Pontia occidentalis occidentalis* (Reakirt), ventral male (Hoosier Pass, Park Co., Colorado).
32a. *Pieris (Artogeia) rapae* (Linnaeus), dorsal male (Broadmoor at Cheyenne Mtn. High School, El Paso Co., Colorado).
32b. *Pieris (Artogeia) rapae* (Linnaeus), dorsal female (Broadmoor, 1/2 mi. N of Cheyenne Mtn. Zoo, El Paso Co., Colorado).
32c. *Pieris (Artogeia) rapae* (Linnaeus), ventral male (Los Angeles, California).
33a. *Euchloe ausonides coloradensis* (Henry Edwards), dorsal male (Phantom Canyon Rd. at Mile Marker 21, Fremont Co., Colorado).
33b. *Euchloe ausonides coloradensis* (Henry Edwards), dorsal female (Phantom Canyon Rd. at Mile Marker 11.5, Fremont Co., Colorado).
33c. *Euchloe ausonides coloradensis* (Henry Edwards), ventral male (Phantom Canyon Rd. at Mile Marker 21, Fremont Co., Colorado).
34a. *Colias philodice eriphyle* W. H. Edwards, dorsal male (Big Spring Ranch, Fish Creek, Teller Co., Colorado).
34b. *Colias philodice eriphyle* W. H. Edwards, dorsal female (Phantom Canyon Rd. at Mile Marker 18 1/4, Fremont Co., Colorado).
34c. *Colias philodice eriphyle* W. H. Edwards, dorsal female, white form (Shoup Rd. Picnic Area, Black Forest, El Paso Co., Colorado).
34d. *Colias philodice eriphyle* W. H. Edwards, ventral male (Florissant, Big Spring Ranch, Teller Co., Colorado).
35a. *Colias eurytheme* Boisduval, dorsal male (San Bernardino Mtns., California).
35b. *Colias eurytheme* Boisduval, dorsal female (San Bernardino Mtns., California).
35c. *Colias eurytheme* Boisduval, dorsal female, white form (Florissant, Big Spring Ranch, Teller Co., Colorado).
35d. *Colias eurytheme* Boisduval, ventral male (Joshua Tree National Monument, California).
36a. *Colias alexandra alexandra* W. H. Edwards, dorsal male (Quick Ranch, SW of Florissant, Teller Co., Colorado).
36b. *Colias alexandra alexandra* W. H. Edwards, dorsal female (Quick Ranch, SW of Florissant, Teller Co., Colorado).
36c. *Colias alexandra alexandra* W. H. Edwards, dorsal female, white form (Quick Ranch, SW of Florissant, Teller Co., Colorado).
36d. *Colias alexandra alexandra* W. H. Edwards, ventral male (Quick Ranch, SW of Florissant, Teller Co., Colorado).
37a. *Eurema nicippe* (Cramer), dorsal male (Fern Forest Co. Park, Broward Co., Florida).
37b. *Eurema nicippe* (Cramer), dorsal female (Parkland, Broward Co., Florida).
37c. *Eurema nicippe* (Cramer), ventral male (Fern Forest Co. Park, Broward Co., Florida).
38a. *Eurema mexicana* (Boisduval), dorsal male (9 mi. W of Rd. 164 on US 287, Childress Co., Texas).
38b. *Eurema mexicana* (Boisduval), dorsal female (Santa Rita Mtns., Madera Canyon, Arizona).
38c. *Eurema mexicana* (Boisduval), ventral male (Acapulco, Summit Chilpancingo, Mexico).
39a. *Nathalis iole* Boisduval, dorsal male (Fern Forest Co. Park, Broward Co., Florida).
39b. *Nathalis iole* Boisduval, dorsal female (1 mi. S of Griffin Rd. on Dykes Rd., Broward Co., Florida).
39c. *Nathalis iole* Boisduval, ventral male (Fern Forest Co. Park, Broward Co., Florida).

COLOR PLATE III

A biologist searching for fossils in the volcanic ash shales at Florissant Fossil Beds National Monument.

A fossil butterfly, *Prodryas persephone*, from the Florissant shales (the type specimen in the Scudder Collection at Harvard University's Museum of Comparative Zoology).

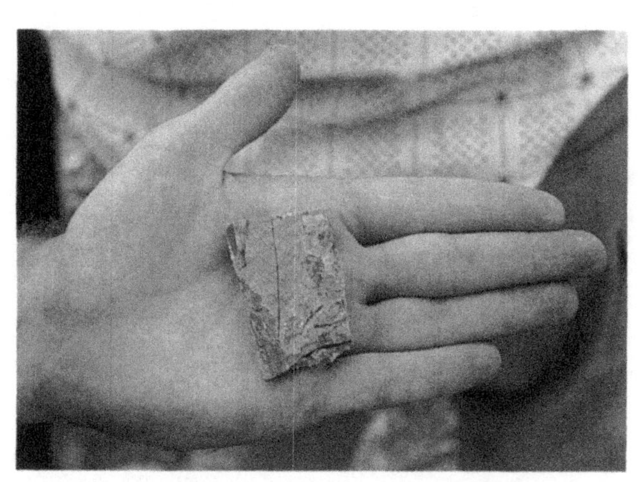

A fossil leaf in the well-preserved condition commonly discovered in the Florissant shales.

Dr. F. Martin Brown, the world's leading expert on Florissant's fossil insects, in the field at the fossil beds.

The type specimen of the fossil butterfly *Oligodonta florissantensis*, uncovered at Florissant in the early 1900's.

COLOR PLATE IV

40a. *Lycaena (Chalceria) rubidus sirius* (W. H. Edwards), dorsal male (Pikes Peak Research Station, Teller Co., Colorado).
40b. *Lycaena (Chalceria) rubidus sirius* (W. H. Edwards), dorsal female (Pikes Peak Research Station, Teller Co., Colorado).
40c. *Lycaena (Chalceria) rubidus sirius* (W. H. Edwards), ventral male (Dome Rock, 7 mi. S of Florissant, Teller Co., Colorado).
41a. *Lycaena (Chalceria) heteronea heteronea* (Boisduval), dorsal male (Parshall, Grand Co., Colorado).
41b. *Lycaena (Chalceria) heteronea heteronea* (Boisduval), dorsal female (Dome Rock, 7 mi. S of Florissant, Teller Co., Colorado).
41c. *Lycaena (Chalceria) heteronea heteronea* (Boisduval), ventral male (Parshall, Grand Co., Colorado).
42a. *Lycaena (Epidemia) helloides* (Boisduval), dorsal male (Wolf Creek Pass, East Slope, Mineral Co., Colorado).
42b. *Lycaena (Epidemia) helloides* (Boisduval), dorsal female (Canon City, Fremont Co., Colorado).
42c. *Lycaena (Epidemia) helloides* (Boisduval), ventral male (Wolf Creek Pass, East Slope, Mineral Co., Colorado).
43a. *Harkenclenus titus titus* (Fabricus), dorsal male (Rock Creek Canyon, S of Colorado Springs, El Paso Co., Colorado).
43b. *Harkenclenus titus titus* (Fabricus), dorsal female (Dome Rock, 7 mi. S of Florissant, Teller Co., Colorado).
43c. *Harkenclenus titus titus* (Fabricus), ventral male (Phantom Canyon Rd. at Mile Marker 11.5, Fremont Co., Colorado).
44a. *Satyrium behrii crossi* (Field), dorsal male (Golden, Lookout Mtn., Colorado).
44b. *Satyrium behrii crossi* (Field), dorsal female (Mt. Herman, Monument Area, El Paso Co., Colorado).
44c. *Satyrium behrii crossi* (Field), ventral male (Solar Trails Co. Park, El Paso Co., Colorado).
45a. *Satyrium californica* (W. H. Edwards), dorsal male (Dome Rock, 7 mi. S of Florissant, Teller Co., Colorado).
45b. *Satyrium californica* (W. H. Edwards), dorsal female (Dome Rock, 7 mi. S of Florissant, Teller Co., Colorado).
45c. *Satyrium californica* (W. H. Edwards), ventral male (Dome Rock, 7 mi. S of Florissant, Teller Co., Colorado).
46a. *Satyrium calanus godarti* (Field), dorsal male (Broadmoor, 1/2 mile N of Cheyenne Mtn. Zoo, El Paso Co., Colorado).
46b. *Satyrium calanus godarti* (Field), dorsal female (Broadmoor, 1/2 mile N of Cheyenne Mtn. Zoo, El Paso Co., Colorado).
46c. *Satyrium calanus godarti* (Field), ventral female (Broadmoor, 1/2 mile N of Cheyenne Mtn. Zoo, El Paso Co., Colorado).
47a. *Callophrys apama homoperplexa* Barnes and Benjamin, dorsal male (Lookout Mtn., Jefferson Co., Colorado).
47b. *Callophrys apama homoperplexa* Barnes and Benjamin, dorsal female (Magnolia Road, Boulder Co., Colorado).
47c. *Callophrys apama homoperplexa* Barnes and Benjamin, ventral male (Broadmoor, 1/2 mi. N of Cheyenne Mtn. Zoo, El Paso Co., Colorado).
48a. *Mitoura spinetorum* (Hewitson), dorsal male (Fraser, Grand Co., Colorado).
48b. *Mitoura spinetorum* (Hewitson), dorsal female (Fraser, Grand Co., Colorado).
48c. *Mitoura spinetorum* (Hewitson), ventral male (Fraser, Grand Co., Colorado).
49a. *Mitoura siva siva* (W. H. Edwards), dorsal male (Magnolia Rd., Boulder Co., Colorado).
49b. *Mitoura siva siva* (W. H. Edwards), dorsal female (Magnolia Rd., Boulder Co., Colorado).
49c. *Mitoura siva siva* (W. H. Edwards), ventral male (Canon City at Red Canon Park, Fremont Co., Colorado).
50a. *Incisalia polios obscurus* Ferris and Fisher, dorsal male (Sugarloaf, Boulder Co., Colorado).
50b. *Incisalia polios obscurus* Ferris and Fisher, dorsal female (Flagstaff Mtn., Boulder Co., Colorado).
50c. *Incisalia polios obscurus* Ferris and Fisher, ventral male (Sugarloaf, Boulder Co., Colorado).
51a. *Incisalia eryphon eryphon* (Boisduval), dorsal male (C.R. 61 at Marigold, Teller Co., Colorado).
51b. *Incisalia eryphon eryphon* (Boisduval), dorsal female (Wild Goat Mtn., SW of Florissant, Park Co., Colorado).
51c. *Incisalia eryphon eryphon* (Boisduval), ventral male (C.R. 61 at Marigold, Teller Co., Colorado).
52a. *Strymon melinus franki* Field, dorsal male (Pikes Peak Research Station, Teller Co., Colorado).
52b. *Strymon melinus franki* Field, dorsal female (Rock Creek Canyon, S of Colorado Springs, El Paso Co., Colorado).
52c. *Strymon melinus franki* Field, ventral male (Wichita Falls, Wichita Co., Texas).
53a. *Hemiargus isola alce* (W. H. Edwards), dorsal male (Quick Ranch, SW of Florissant, Teller Co., Colorado).

COLOR PLATE V

61a. *Agraulis vanillae incarnata* (Riley), dorsal male (San Antonio, Bexar Co., Texas).
61b. *Agraulis vanillae incarnata* (Riley), dorsal female (San Antonio, Bexar Co., Texas).
61c. *Agraulis vanillae incarnata* (Riley), ventral male (San Antonio, Bexar Co., Texas).
62a. *Polygonia interrogationis* (Fabricus), dorsal female, short-day, spring or fall form (Smithtown, New York).
62b. *Polygonia interrogationis* (Fabricus), dorsal female, long-day, summer form (Smithtown, New York).
62c. *Polygonia interrogationis* (Fabricus), ventral female (Smithtown, New York).
68a. *Vanessa virginiensis* (Drury), dorsal male (Norman, Oklahoma).
68b. *Vanessa virginiensis* (Drury), dorsal female (Punky Swamp, Kennebunk, Maine).
68c. *Vanessa virginiensis* (Drury), ventral male (Haydenville, Massachusetts).
69a. *Vanessa cardui* (Linnaeus), dorsal male (Jackson, Independence Co., Missouri).
69b. *Vanessa cardui* (Linnaeus), dorsal female (Spring Creek, Benner Twp., Centre Co., Pennsylvania).
69c. *Vanessa cardui* (Linnaeus), ventral male (Phantom Canyon Rd. at Mile Marker 11.5, Fremont Co., Colorado).
70a. *Vanessa atalanta rubria* (Fruhstorfer), dorsal male (Pine Island Ridge, Davie, Broward Co., Florida).
70b. *Vanessa atalanta rubria* (Fruhstorfer), dorsal female (4.2 mi. NE of Abita Springs, St. Tam. Parish, Louisiana).
70c. *Vanessa atalanta rubria* (Fruhstorfer), ventral male (Pine Island Ridge, Davie, Broward Co., Florida).
71a. *Euptoieta claudia* (Cramer), dorsal male (Phantom Canyon Rd. at Mile Marker 21, Fremont Co., Colorado).
71b. *Euptoieta claudia* (Cramer), dorsal female (Pikes Peak Research Station, Teller Co., Colorado).
71c. *Euptoieta claudia* (Cramer), ventral male (Pikes Peak Research Station, Teller Co., Colorado).

Dorsal and ventral surfaces of a male Orange-Margined Blue, *Lycaeides melissa melissa*, at the Fossil Beds National Monument.

An Ochre Ringlet, *Coenonympha ochracea*, feeding on a fleabane flower at Florissant.

The Chryxus Arctic, *Oeneis chryxus*, lands momentarily on a weathering log at 12,000 feet in Leavick Valley.

The Dark Wood Nymph, *Cercyonis oetus charon*, feeding on Yarrow (*Achillea*) at Florissant.

The Common White, *Pieris protodice*, which becomes fairly abundant during the summer in the Florissant region and feeds on many composite species (here on *Senecio* at Florissant).

The Dark Wood Nymph, *Cercyonis oetus charon*, feeding on nectar from *Senecio* flowers at Florissant. This male has been given a black mark on the forewing as part of a mark-release study to determine population size and adult movements.

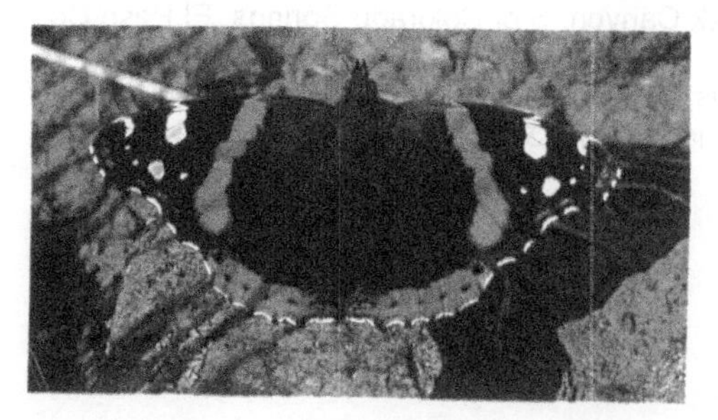

The Red Admiral, *Vanessa atalanta*, sunning itself on the ground in Park County, Colorado.

COLOR PLATE V

COLOR PLATE VI

63a. *Polygonia satyrus* (W. H. Edwards), dorsal male (Phantom Canyon Rd. at Mile Marker 9, Fremont Co., Colorado).
63b. *Polygonia satyrus* (W. H. Edwards), dorsal female (Phantom Canyon Rd. at Mile marker 21, Fremont Co., Colorado).
63c. *Polygonia satyrus* (W. H. Edwards), ventral male (C.R. 61 at Marigold, Teller Co., Colorado).
64a. *Polygonia faunus hylas* (W. H. Edwards), dorsal male (Bigelow Divide, San Isabel Forest, Custer Co., Colorado).
64b. *Polygonia faunus hylas* (W. H. Edwards), dorsal female (Mt. Flagstaff, Boulder, Colorado).
64c. *Polygonia faunus hylas* (W. H. Edwards), ventral male (Bigelow Divide, San Isabel Forest, Custer Co., Colorado).
65a. *Polygonia zephyrus* (W. H. Edwards), dorsal male (Phantom Canyon Rd. at Mile Marker 11.5, Fremont Co., Colorado).
65b. *Polygonia zephyrus* (W. H. Edwards), dorsal female (Pikes Peak Research Station, Teller Co., Colorado).
65c. *Polygonia zephyrus* (W. H. Edwards), ventral male (4 mi. S of Cripple Creek on C.R. 61, Teller Co., Colorado).
77a. *Phyciodes tharos pascoensis* Wright, dorsal male (Dome Rock, 7 mi. S of Florissant, Teller Co., Colorado).
77b. *Phyciodes tharos pascoensis* Wright, dorsal female (Broadmoor, 1/2 mi. N of Cheyenne Mtn. Zoo, El Paso Co., Colorado).
77c. *Phyciodes tharos pascoensis* Wright, ventral male (Pikes Peak Research Station, Teller Co., Colorado).
78a. *Phyciodes campestris camillus* W. H. Edwards, dorsal male (Bear Creek Regional Park, El Paso Co., Colorado).
78b. *Phyciodes campestris camillus* W. H. Edwards, dorsal female (Bear Creek Regional Park, El Paso Co., Colorado).
78c. *Phyciodes campestris camillus* W. H. Edwards, ventral male (Pikes Peak Research Station, Teller Co., Colorado).
79a. *Phyciodes vesta* (W. H. Edwards), dorsal male (Hondo, Medina Co., Texas).
79b. *Phyciodes vesta* (W. H. Edwards), dorsal female (Hondo, Medina Co., Texas).
79c. *Phyciodes vesta* (W. H. Edwards), ventral male (Hondo, Medina Co., Texas).
81a. *Chlosyne (Charidryas) gorgone carlota* (Reakirt), dorsal male (C.R. 61 at Marigold, Teller Co., Colorado).
81b. *Chlosyne (Charidryas) gorgone carlota* (Reakirt), dorsal female (Phantom Canyon Rd. at Jct. with C.R. 123, Fremont Co., Colorado)
81c. *Chlosyne (Charidryas) gorgone carlota* (Reakirt), ventral male (Rock Creek Canyon, S of Colorado Springs, El Paso Co., Colorado).
82a. *Chlosyne palla calydon* (Holland), dorsal male (Aspen, Pitkin Co., Colorado).
82b. *Chlosyne palla calydon* (Holland), dorsal female (Fraser, Grand Co., Colorado).
82c. *Chlosyne palla calydon* (Holland), ventral male (Aspen, Pitkin Co., Colorado).
83a. *Thessalia fulvia* (W. H. Edwards), dorsal male (Royal Gorge Picnic Area, Fremont Co., Colorado).
83b. *Thessalia fulvia* (W. H. Edwards), dorsal male (Trail of Serpent, Grand Junction, Colorado).
83c. *Thessalia fulvia* (W. H. Edwards), ventral male (Trail of Serpent, Colorado National Monument, Mesa Co., Colorado).
84a. *Poladryas arachne arachne* (W. H. Edwards), dorsal male (Dome Rock, 7 mi. S of Florissant, Teller Co., Colorado).
84b. *Poladryas arachne arachne* (W. H. Edwards), dorsal female (Quick Ranch, SW of Florissant, Teller Co., Colorado).
84c. *Poladryas arachne arachne* (W. H. Edwards), ventral male (Big Spring Ranch, Florissant, Teller Co., Colorado).
85a. *Euphydryas (Occidryas) anicia capella* (Barnes), dorsal male (Pikes Peak Research Station, Teller Co., Colorado).
85b. *Euphydryas (Occidryas) anicia capella* (Barnes), dorsal female (Quick Ranch, SW of Florissant, Teller Co., Colorado).
85c. *Euphydryas (Occidryas) anicia capella* (Barnes), ventral male (Quick Ranch, SW of Florissant, Teller Co., Colorado).

COLOR PLATE VI

COLOR PLATE VII

66a. *Nymphalis antiopa antiopa* (Linnaeus), dorsal female (Pikes Peak Research Station, Teller Co., Colorado).
66b. *Nymphalis antiopa antiopa* (Linnaeus), ventral male (Quanah at Georgia-Pacific Plant, Hardeman Co., Texas).
67a. *Nymphalis (Aglais) milberti milberti* Godart, dorsal male (Pikes Peak Research Station, Teller Co., Colorado).
67b. *Nymphalis (Aglais) milberti milberti* Godart, ventral male (Phantom Canyon Rd. at Mile Marker 11.5, Fremont Co., Colorado).
72a. *Speyeria idalia* (Drury), dorsal male (Sterling, Kansas).
72b. *Speyeria idalia* (Drury), dorsal female (Big Spring Ranch, Park Co., Colorado).
72c. *Speyeria idalia* (Drury), ventral male (Chippewa Hills, Franklin Co., Kansas).
73a. *Speyeria edwardsii* (Reakirt), dorsal male (Running Creek Field Station, Elbert Co., Colorado).
73b. *Speyeria edwardsii* (Reakirt), dorsal female (Flagstaff Mtn., Boulder Co., Colorado).
73c. *Speyeria edwardsii* (Reakirt), ventral male (Foothills Highway, Boulder Co., Colorado).
74a. *Speyeria coronis halcyone* (W. H. Edwards), dorsal male (Boulder Canyon, Boulder Co., Colorado).
74b. *Speyeria coronis halcyone* (W. H. Edwards), dorsal female (Quick Ranch, SW of Florissant, Teller Co., Colorado).
74c. *Speyeria coronis halcyone* (W. H. Edwards), ventral male (Bluebell Canyon, Boulder Co., Colorado).
75a. *Speyeria atlantis electa* (W. H. Edwards) or *hesperis* (W. H. Edwards), dorsal male (Phantom Canyon Rd. at Mile Marker 11.5, Fremont Co., Colorado).
75b. *Speyeria atlantis electa* (W. H. Edwards) or *hesperis* (W. H. Edwards), dorsal female (1 mile SE of Golden, Jefferson Co., Colorado).
75c. *Speyeria atlantis electa* (W. H. Edwards) or *hesperis* (W. H. Edwards), ventral male, unsilvered form (Phantom Canyon Rd. at Mile Marker 11.5, Fremont Co., Colorado).
75d. *Speyeria atlantis electa* (W. H. Edwards) or *hesperis* (W. H. Edwards), ventral male, silvered form (Phantom Canyon Rd. at Mile Marker 11.5, Fremont Co., Colorado).
76a. *Speyeria mormonia eurynome* (W. H. Edwards), dorsal male (Horseshoe Mtn., W of Fairplay, Park Co., Colorado).
76b. *Speyeria mormonia eurynome* (W. H. Edwards), dorsal female (Horseshoe Mtn., W of Fairplay, Park Co., Colorado).
76c. *Speyeria mormonia eurynome* (W. H. Edwards), ventral male (Horseshoe Mtn., W of Fairplay, Park Co., Colorado).
86a. *Limenitis (Basilarchia) weidemeyerii weidemeyerii* (W. H. Edwards), dorsal male (Top of the World, SW of Florissant, Teller Co., Colorado).
86b. *Limenitis (Basilarchia) weidemeyerii weidemeyerii* (W. H. Edwards), ventral male (Phantom Canyon Rd. at Mile Marker 11.5, Fremont Co., Colorado).

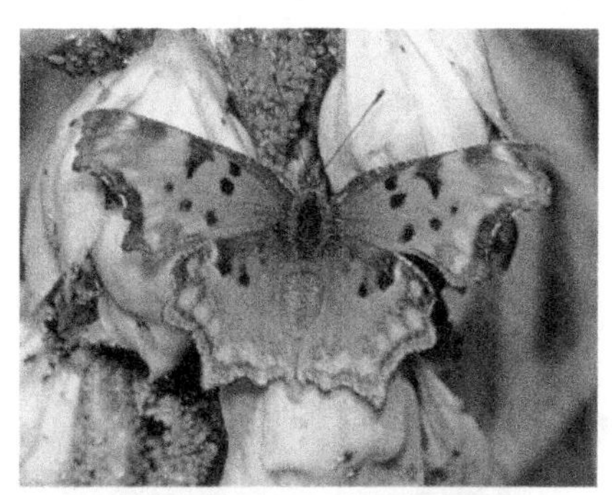

The Zephyr Anglewing, *Polygonia zephyrus*, feeding on aphis exudates on *Yucca* flowers near Cripple Creek, Colorado.

The Satyr Anglewing, *Polygonius satyrus* (ventral side), feeding on a thistle flower at Florissant.

The communal larvae of Milbert's Tortoiseshell, *Nymphalis milberti*, feeding on stinging nettles at Fossil Beds National Monument.

COLOR PLATE VII

COLOR PLATE VIII

87a. *Cyllopsis pertepida dorothea* (Nabokov), dorsal male (Broadmoor, 1/2 mile N of Cheyenne Mtn. Zoo, El Paso Co., Colorado).
87b. *Cyllopsis pertepida dorothea* (Nabokov), dorsal female (4 mi. S of Cripple Creek on C.R. 61, Teller Co., Colorado).
87c. *Cyllopsis pertepida dorothea* (Nabokov), ventral male (Broadmoor at Cheyenne Mtn. High School, El Paso Co., Colorado).
88a. *Coenonympha tullia ochracea* W. H. Edwards, dorsal male (Pikes Peak Research Station, Teller Co., Colorado).
88b. *Coenonympha tullia ochracea* W. H. Edwards, dorsal female (Horseshoe Mtn., W of Fairplay, Park Co., Colorado).
88c. *Coenonympha tullia ochracea* W. H. Edwards, ventral male (Phantom Canyon Rd. at Mile Marker 18 1/4, Fremont Co., Colorado).
89a. *Cercyonis meadii meadii* (W. H. Edwards), dorsal male (Broadmoor at Cheyenne Mtn. High School, El Paso Co., Colorado).
89b. *Cercyonis meadii meadii* (W. H. Edwards), dorsal female (Broadmoor at Cheyenne Mtn. High School, El Paso Co., Colorado).
89c. *Cercyonis meadii meadii* (W. H. Edwards), ventral male (Lone Rock Camp at Deckers, Douglas Co., Colorado).
90a. *Cercyonis oetus charon* (W. H. Edwards), dorsal male (Pikes Peak Research Station, Teller Co., Colorado).
90b. *Cercyonis oetus charon* (W. H. Edwards), dorsal female (Big Spring Ranch, Florissant, Teller Co., Colorado).
90c. *Cercyonis oetus charon* (W. H. Edwards), ventral male (Quick Ranch, S of Florissant, Teller Co., Colorado).
91a. *Erebia epipsodea epipsodea* Butler, dorsal male (Horseshoe Mtn, W. of Fairplay, Park Co., Colorado).
91b. *Erebia epipsodea epipsodea* Butler, dorsal female (Horseshoe Mtn, W of Fairplay, Park Co., Colorado).
91c. *Erebia epipsodea epipsodea* Butler, ventral male (Dome Rock, 7 mi. S of Florissant, Teller Co., Colorado).
92a. *Neominois ridingsii ridingsii* (W. H. Edwards), dorsal male (4 mi. S of Cripple Creek on C.R. 61, Teller Co., Colorado).
92b. *Neominois ridingsii ridingsii* (W. H. Edwards), dorsal female (Dome Rock, 7 mi. S of Florissant, Teller Co., Colorado).
92c. *Neominois ridingsii ridingsii* (W. H. Edwards), ventral male (Dome Rock, 7 mi. S of Florissant, Teller Co., Colorado).
93a. *Oeneis chryxus chryxus* (Doubleday and Hewitson), dorsal male (Horseshoe Mtn., W of Fairplay, Park Co., Colorado).
93b. *Oeneis chryxus chryxus* (Doubleday and Hewitson), dorsal female (Horseshoe Mtn., W of Fairplay, Park Co., Colorado).
93c. *Oeneis chryxus chryxus* (Doubleday and Hewitson), ventral male (Horseshoe Mtn., W of Fairplay, Park Co., Colorado).
94a. *Oeneis uhleri uhleri* (Reakirt), dorsal male (Pikes Peak Research Station, Teller Co., Colorado).
94b. *Oeneis uhleri uhleri* (Reakirt), dorsal female (Top of the World, SW of Florissant, Teller Co., Colorado).
94c. *Oeneis uhleri uhleri* (Reakirt), ventral male (Top of the World, SW of Florissant, Teller Co., Colorado).
95a. *Oeneis alberta oslari* Skinner, dorsal male (Crook Creek, South Park, Park Co., Colorado).
95b. *Oeneis alberta oslari* Skinner, dorsal female (Crook Creek, South Park, Park Co., Colorado).
95c. *Oeneis alberta oslari* Skinner, ventral male (Crook Creek, South Park, Park Co., Colorado).

COLOR PLATE VIII

COLOR PLATE IX

27a. *Battus philenor philenor* (Linnaeus), dorsal male (Big Spring Ranch, SW of Florissant, Teller Co., Colorado).
27b. *Battus philenor philenor* (Linnaeus), ventral male (Big Spring Ranch, SW of Florissant, Teller Co., Colorado).
80a. *Phyciodes pallidus* (W. H. Edwards), dorsal female (Big Spring Ranch, SW of Florissant, Teller Co., Colorado).
80b. *Phyciodes pallidus* (W. H. Edwards), ventral female (Big Spring Ranch, SW of Florissant Teller Co., Colorado).
97a. *Danaus gilippus* (Cramer), dorsal male (Nature Place, SW of Florissant, Teller Co., Colorado).

COLOR PLATE IX

The wings are mostly very pale yellow with some orange basally. Males also have an orange patch across the costal margin of the upper hindwings. The Mexican Yellow rarely strays into the Florissant area from lower elevations. This species may temporarily colonize areas of Colorado where locust trees *(Robinia neomexicana)* (Fabaceae), a larval host (Scott 1986), have been planted.

39. *Nathalis iole* Boisduval DWARF SULPHUR

Plate III, Fig. 39a-c

The Dwarf Sulphur is the smallest pierid butterfly in our area. Males are yellow and black. Females are darker yellow above and are green below. Despite its small size, the Dwarf Sulphur is a renowned migrant, that each year colonizes areas far beyond its overwintering range. A few individuals of this tiny pierid have been found in dry meadows in the Florissant Valley during July and August.

Females lay yellow eggs singly on certain composites (Asteraceae) (especially seedlings and water-stressed individuals). It is not known if the Dwarf Sulphur actually breeds successfully from year to year in montane areas of Colorado.

Family LYCAENIDAE
THE HAIRSTREAKS, COPPERS, and BLUES

The Lycaenidae occur throughout the world, even in the Arctic, and number about 4,089 species (Shields, 1989b). Compared with the members of the two preceding families, they are much smaller in size and include in their embrace the smallest butterfly in the world. There are 122 species in North America, of which 20 have been recorded from the Florissant region. Only four of the eight subfamilies (treated as tribes by some authorities) occur in North America, of which three are represented in our fauna: the Lycaeninae (Coppers) with three species, the Eumaeninae (Hairstreaks) with ten species, and the Polyommatinae (Blues) with seven species. Most lycaenids are local in distribution and do not migrate, but many have a strong flight. The forelegs of males differ from the four mid and hind legs by being slightly shorter, with fused tips and no claws. In females, however, all six legs are of the same length and all bear claws. Flight habits and basking behavior vary considerably in this group, but most flights are short and most species rest with their wings closed. A peculiar behavior of resting lycaenids, especially hairstreaks, is hindwing rubbing, in which the hindwings are moved back and forth in opposite directions. Such behavior may be adaptive by drawing attention to the eyespots and antenna-like tails on the hindwing, distracting a predator's interest away from the less-disposable business end of the butterfly. Many species of dicotyledonous plants are eaten by the larvae, which often specialize on the flowers and young fruits. Larvae are slug-like and some secrete sweet fluids from special dorsal abdomenal glands. The latter species attract symbiotic ants to this fluid and the ants protect them against parasitic wasps and other enemies.

Subfamily LYCAENINAE

40. *Lycaena (Chalceria) rubidus sirius* (W. H. Edwards) Plate IV, Fig. 40a-c
RUDDY COPPER

The males of the Ruddy Copper are brilliant orange on the dorsal surface and whitish on the ventral surface. This is one of the most distinctive and attractive butterflies in the Florissant region. The females are highly variable in color, ranging from almost as light an orange as the males to a patchy gray and brown dorsal surface to being almost entirely melanic. The males begin flying around the end of June, while the females reach peak abundance in the last ten days of July and early August. This species is restricted to moist meadows, usually occurring along the semi-permanent streams where tall coarse grasses occur. They land frequently on flowers of Shrubby Cinquefoil *(Pentaphylloides floribunda)* and on the stems of the tall grasses or dock (*Rumex*).

FIGURE 36. A female Ruddy Copper (*Lycaena rubidus*) sips nectar from a composite flower.

The larval host plant is Willow Dock *(Rumex salicifolius* ssp. *triangulivalvis)* (Polygonaceae) The white eggs are laid by the females on both living and dead *Rumex* plants and in the debris beneath them on the ground. Fall diapause and winter hibernation occur in the egg stage. Mature larvae are brown in ground color, with a reddish band down the dorsum (Scott 1986). Like most lycaenids, Ruddy Copper larvae are stout-bodied and slow-moving. The head is kept retracted into the first thoracic segment but can be extended while feeding. The pupae are attached to the substrate by a silk thread around the thorax and by hooks at the tip of the abdomen.

Scores of these butterflies occur along streambeds and the males, though rapid fliers, are seemingly territorial, defending small areas around their perching sites. Both sexes avidly visit flowers, especially composites (Figure 36) and Shrubby Cinquefoil. Territorial behavior of the Ruddy Copper would be an extremely interesting subject for study.

41. *Lycaena (Chalceria) heteronea heteronea* (Boisduval) Plate IV, Fig. 41a-c
BLUE COPPER

Males of the Blue Copper look very similar to species in the subfamily Polyommatinae (the Blues). This is the only copper in North America with this brilliant blue coloration. The females are gray or grayish brown on the dorsal surface, with a flush of blue. The Blue Copper flies in association with various wild buckwheats (*Eriogonum* species), which are patchily distributed in the Florissant region. Small colonies of the Blue Copper exist in association with Sulphur-Flower *(Eriogonum umbellatum)* (Polygonaceae) in the lower part of the Mueller Ranch area near Dome Rock. Adults readily visit *Eriogonum* species to sip nectar and lay eggs.

Feeding on buckwheat leaves, the mature larva is slug-like in shape, dull greenish-gray in ground color, and thickly covered with white pile and a scattering of low white tubercles. The pupa is mottled green, with a green dorsal line and lateral markings of green. The Blue Copper is normally found in rather open sagebrush flats and slopes from 6,200 to 9,800' elevation in Colorado. Overwintering occurs in the egg stage.

42. *Lycaena (Epidemia) helloides* (Boisduval) Plate IV, Fig. 42a-c
PURPLISH COPPER

The Purplish Copper is basically a low-elevation species found on the prairies and in the foothills of the Rockies, usually below 8,200' elevation. This species is found rarely during July in the area immediately around Florissant and the Fossil Beds National Monument. It typically occurs along streams where the males perch on vegetation and defend small territories while awaiting females.

The larval foodplants are in the family Polygonaceae and include species of Knotweed *(Polygonum)* and Dock *(Rumex)*. Eggs are laid on the foodplant flowers and seeds. Overwintering occurs in the egg stage. The mature larva is grass-green with numerous tubercles, each tubercle covered with pale hairs and a spine. There are two dorsal and two lateral lines or stripes of yellow, with many diagonal lines on the sides. Pupation occurs in the debris at the base of the foodplant. Adults are capable of moving long distances, and the species is known to have two to three broods through the summer at other locations in the Rocky Mountains.

Subfamily EUMAENINAE

43. ***Harkenclenus titus titus*** **(Fabricius)** Plate IV, Fig. 43a-c
CORAL HAIRSTREAK

The Coral Hairstreak lacks tails characteristic of many hairstreaks, and is brownish-gray on the upperside. The underside of the hindwings has a distinct band of red spots along the outer margin. Adult males bear a small oval patch of androconial (scent) scales midway along the costa of the forewings. The nominate subspecies is found throughout the eastern United States and southeastern Canada, west to east-central Colorado. The Coral Hairstreak is relatively common in foothill canyons but is rare in the Florissant region. Adults visit the flowers of dogbane (*Apocynum*), Shrubby Cinquefoil (*Pentaphylloides floribunda*), and various composites.

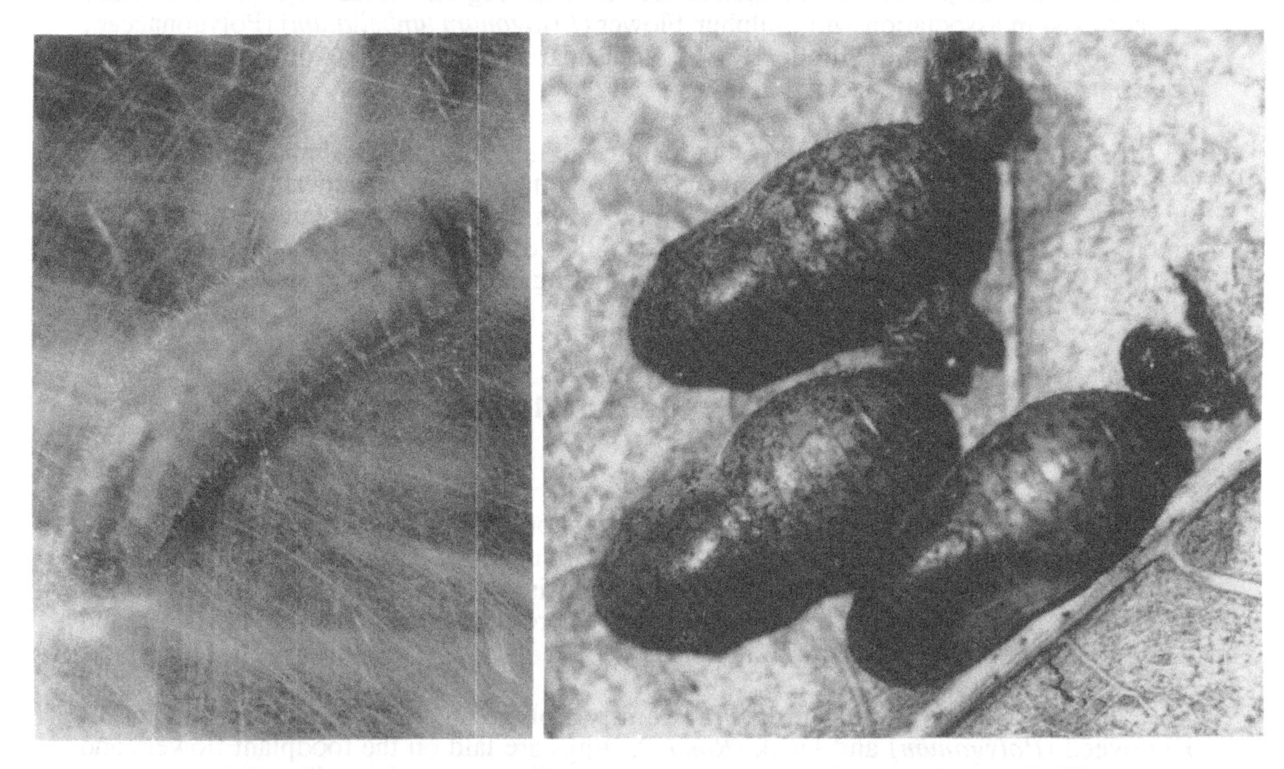

FIGURE 37. *Harkenclenus titus titus* (Species 43), last instar larva (left) and pupae (right) from 8,800' elevation in Teller County, Colorado.

The specimens taken here may represent stray individuals rather than a resident population, since few woody Rosaceae (Wild Cherry or Wild Plum, *Prunus*), the reported larval foodplants, occur in the Florissant area. The life-history stages illustrated in Figure 37 were reared from eggs produced by a female taken by M. C. Minno at 8800' in Teller County. Eggs are laid singly on the branches of the host plant. The mature larva is yellowish-green with rose-colored patches on the thorax and caudal end. The stout pupa is dark brown, mottled with black. The Coral Hairstreak overwinters in the egg stage.

44. *Satyrium behrii crossi* (Field) BEHR'S HAIRSTREAK

Plate IV, Fig. 44a-c

Behr's Hairstreak is distinguished by the bright orange areas in the center of each of the wings dorsally and the wide black-brown margins. It is also tailless. The few specimens that have been taken near Florissant are apparently immigrants from lower elevations; Behr's Hairstreak occurs more abundantly in the juniper scrub zone. Adults perch on shrubs and avidly visit flowers, such as wild buckwheat (*Eriogonum*).

The larval foodplants are Bitterbrush or Antelope Brush (*Purshia tridentata*) and Mountain Mahogany (*Cercocarpus montanus*) (both Rosaceae). The larvae are greenish in ground color, with white and yellow mottling across the body surface, rendering them well-camouflaged on the foodplant. The pupae are brown. Adults fly from mid-June to July. Overwintering is in the egg stage.

45. *Satyrium californica* (W. H. Edwards) CALIFORNIA HAIRSTREAK

Plate IV, Fig. 45a-c

This species is generally similar in appearance to the preceding two, but the ground color above is a golden brown rather than a dark brown and on the underside the submarginal band on the hindwing is composed of well-separated, large black spots (Figure 38). The presence of a thin tail and more rounded anal margin will differentiate it from the Coral Hairstreak *(H. titus)*. The California Hairstreak flies in one brood from late June to July. Adults avidly visit the flowers of *Asclepias speciosa* (milkweed) and perch on shrubs. A small population of the California Hairstreak, associated with Mountain Mahogany *(Cercocarpus montanus)*, occurs on an open dry hillside near Dome Rock.

FIGURE 38. A California Hairstreak (*Satyrium californica)* sips nectar from a *Senecio* flower at the Pikes Peak Research Station.

The eggs are laid in groups at the base of the leaves of Mountain Mahogany *(Cercocarpus montanus*). Wild buckwheat (*Eriogonum*) may also be used (Scott, 1986). The mature larva is grayish brown, with white or gray spots along the mid-dorsal line. The eggs overwinter.

46. ***Satyrium calanus godarti*** **(Field)** Plate IV, Fig. 46a-c
BANDED HAIRSTREAK

The type locality for this subspecies is Rosemont (a location east of Cripple Creek in the southeastern corner of Teller County, next to El Paso County). The Banded Hairstreak is very similar in appearance to the Coral Hairstreak *(Harkenclenus titus)*; however, the latter is tailless and has a more triangular hindwing, whereas *S. c. godarti* has tails. Also, the Banded Hairstreak has streak-like whitish highlighting along the black markings on the ventral forewing and hindwing, whereas on the Coral Hairstreak, the white highlighting surrounds black spot-like markings. The latter characteristic should distinguish specimens that have the outer anal angle of the hindwing broken or otherwise damaged. This species strays into the Florissant region on occasion, and may occur together with *H. titus*.

The Banded Hairstreak is always found in association with oaks, primarily Gambel's Oak *(Quercus gambelii)* (Fagaceae). It has also been reported in association with *Prunus virginiana* (Wild Cherry) in eastern Douglas County, Colorado. The single brood flies from late June to July. Last instar larvae may be either light green with two faint white lines on the dorsum, or light gray with the white markings outlined in red. Pupae are brown. Overwintering occurs in the egg stage.

47. ***Callophrys apama homoperplexa*** **Barnes & Benjamin** Plate IV, Fig. 47a-c
CANYON GREEN HAIRSTREAK

The Canyon Green Hairstreak is dark brown on the upperside and a vivid grass green on the ventral hindwing surface. There is a variable row of white post-median spots zig-zagging across the ventral hindwing surface. The sexes are similar, but females tend to have more orange dorsally and males have noticeable scent-scale patches on the forewing. This tailless hairstreak is encountered in open pine forests and in small valleys within the forest. It is always uncommon. Adults may be taken throughout the month of June, and occasionally worn adults are taken in July.

The reported Rocky Mountain foodplant for this species, Sulphur Flower (*Eriogonum umbellatum*), occurs only at Dome Rock in our area. The Canyon Green Hairstreak may also use Winged Buckwheat *(Eriogonum alatum)*, the only wild buckwheat species to be commonly found in this area. No detailed published descriptions on the life history of this species are available. However, the larvae are reported to eat *Eriogonum* flowers and occur in either green or red color phases (Ferris and Brown 1981). Overwintering occurs in the pupal stage.

48. *Mitoura spinetorum* (Hewitson) — Plate IV, 48a-c
THICKET HAIRSTREAK

The Thicket Hairstreak is usually a rare hairstreak everywhere it occurs in the West. In the Florissant region it flies during late June around pine trees infested with dwarf mistletoe, the larval host plant. The tailed adults may be recognized immediately by the steely blue-gray upperside and the distinctive rich dark chestnut brown underside bearing a continuous white line on both forewing and hindwing.

The larval foodplants include various species of dwarf mistletoes in the genus *Arceuthobium* (Viscaceae), which are parasitic on pine trees. *Arceuthobium vaginatum* occurs frequently on Ponderosa Pine in the Florissant area. The mature larva is olive yellow with a somewhat lighter mid-dorsal stripe and dull white markings on each segment that break up the appearance and make it blend very well with the yellowish host plant stems. Pupation occurs in mistletoe clusters on pine boughs. The pupae are brown in color. Adults emerge from diapausing pupae in late June and early July. The species occurs in very low numbers and collecting several adults in one year would be considered a normal catch.

49. *Mitoura siva siva* (W. H. Edwards) — Plate IV, Fig. 49a-c
JUNIPER HAIRSTREAK

The Juniper Hairstreak is a somewhat smaller species than the preceding and has a green ventral surface with a jagged white line across the hindwing. It is normally an inhabitant of the lower zones where juniper occurs and the specimens collected in the Florissant region are apparently strays from lower elevations to the south. It rarely occurs above 7,000' in elevation. Adults fly from late May to July, and commonly perch near the tops of juniper trees. They readily visit yellow composite flowers for nectar.

The larval foodplants are junipers *(Juniperus)* (Cupressaceae). The eggs are laid in the terminal twig bracts and the resulting larvae complete development in one season. The mature larva is greenish in ground color with lemon yellow chevrons on either side of the mid-dorsal region of each segment. Pupae are brown. Hibernation is in the pupal stage. The nearest suitable host junipers (other than the prostrate Common Juniper, *Juniperus communis*, which occurs in the Florissant region but does not seem to be used are located about two miles to the west of Pikes Peak Research Station on rocky, south-facing slopes north of Slater Creek.

50. *Incisalia polios obscurus* Ferris & Fisher — Plate IV, Fig. 50a-c
HOARY ELFIN

The Hoary Elfin has a distinctive gray suffusion across the outer half of the ventral hindwing surface. It occurs in open Ponderosa Pine forest but is relatively rare and occurs only in low numbers. In our area it has been recorded only in mid-May.

The larva feeds on Bearberry or Kinnikinnik *(Arctostaphylos uva-ursi)* (Ericaceae), which grows as an evergreen ground cover in open coniferous forest. The larva is reported to be green and the pupa is brown, but the life history has not been completely described.

51. ***Incisalia eryphon eryphon*** **(Boisduval)** Plate IV, Fig. 51a-c
WESTERN PINE ELFIN

The Western Pine Elfin is easily recognized by its nut-brown dorsal surface, boldly checkered underside, and white wing fringes. The underside of the hindwing is purplish-brown in ground color. The adults fly in Ponderosa Pine forest areas in May to early June, particularly around the edges of forest glades. They occasionally sip water from muddy stream banks.

The larval host is Ponderosa Pine (*Pinus ponderosa*) and the eggs are laid near the growing tips of the pine branches. The mature larva is velvety green with creamy white markings to either side of the mid-dorsal area on each segment, as well as around the spiracles. The body is covered by tan hairs. The pupa is brown. The species flies in one brood and overwinters in the pupal stage.

52. ***Strymon melinus franki*** **Field** Plate IV, Fig. 52a-c
GRAY HAIRSTREAK

The Gray Hairstreak is a common lycaenid butterfly across North America. Its status as a breeding resident in our area is uncertain; scattered individuals are taken in dry meadows during July and August and they are almost always worn specimens, probably indicating a long period of flight before arrival in our area. Adults frequent meadows where they visit flowers such as White Dutch Clover, and perch on vegetation. This species often rests in a head-downward position. The ground color on the dorsal surface is slate gray and at the base of the hindwing there is a prominent red-orange eyespot with a black pupil. The hindwings bear short, thin tails. The two sexes can be distinguished easily by the fact that the abdomen of the male is orange while that of the female is the same gray color as the wing. Unlike other hairstreaks in the Florissant region, males lack scent-scale patches on the forewing.

It is possible for this species to be a breeding resident of our area, for its many larval foodplants include various local legume genera, especially milk vetch *(Astragalus)* and mallows (Malvaceae). The larvae feed inside buds and fruits of these plants. The mature larva is green or reddish-brown in ground color and is covered with short tan hairs. Pupae are light brown with dark markings. This species overwinters in the pupal stage in the lowlands, but while it regularly colonizes mountain locations, it does not appear to be able to overwinter at high altitudes.

Subfamily POLYOMMATINAE

53. *Hemiargus isola alce* (W. H. Edwards) Plate IV, Fig. 53a-c
REAKIRT'S BLUE

Reakirt's Blue is a distinctive small species with purplish-blue dorsal surface and large black spots on the ventral surface of the forewing and the inner part of the hindwing. The females are mostly dark brown dorsally. The species is widespread in the western United States and occurs from the dry lowlands up to about 10,000'. It is moderately common in the Florissant region, although it has a rather weak flight close to the ground and may be easily overlooked.

The larval foodplants include several genera of legumes found in our area, including milk vetch *(Astragalus)* and Sweet Clover (*Melilotis officinalis*) (Fabaceae). This species probably does not overwinter successfully north of Colorado, but may become temporarily established by annual summer migration. It is also doubtful that Reakirt's Blue overwinters at our elevations. However, it may go through two broods here, the first in June and early July, and the second in August.

54. *Everes amyntula amyntula* (Boisduval) Plate IV, Fig. 54a-c
WESTERN TAILED BLUE

The Western Tailed Blue occurs from the foothills of the Rockies westward to the Pacific Coast. The Eastern Tailed Blue *(Everes comyntas* Godart) occurs up to the foothills near Denver in eastern Colorado and from there eastward to the Atlantic Coast. Our material from the Florissant region seems closest to *E. amyntula.* The most distinguishing features of this Blue from our other species are the rounded forewing tips and the tiny tails at the anal margin where well-developed orange lunules cap the dark dot at the base of the tail. This butterfly prefers wet meadow habitats and occasionally strays into adjacent dry meadows. Individuals are never abundant but are found from mid-June through July in the Florissant area.

The larval foodplants include various Fabaceae, such as Milk Vetch *(Astragulus)*; the larvae feed on the young seeds inside the seedpods. Other hosts include Loco-weeds *(Oxytropis)* (Fabaceae) and clovers *(Trifolium).* The mature larva has a variable green ground color, with pink or maroon markings and a covering of short white hairs. Nearly mature larvae hibernate.

55. *Celastrina ladon cinerea* (W. H. Edwards) Plate IV, Fig. 55a-c
SPRING AZURE

This is one of the first butterflies of the spring and is appropriately named the Spring Azure. The males are light, shining, lilac blue dorsally, bordered with a very fine black line and an outer black-and-white checkered fringe. The outer third of the forewing of the

FIGURE 39. *Celastrina ladon cinerea* (Species 55), last instar larva and pupa (Teller County, Colorado) on *Jamesia americana* flowers.

FIGURE 40. Last instar larva of *Glaucopsyche lygdamus oro* (Species 56) from Teller County, Colorado, feeding on flowers of *Oxytropis*.

female is suffused with brownish-black. Adults are on the wing from late May through mid-June and there does not appear to be a second brood.

The larval foodplants include Wild Cherry (*Prunus virginiana*), Sticky Laurel (*Ceanothus velutinus*) (Rhamnaceae), and Waxflower (*Jamesia americana*) (Hydrangeaceae). The larvae are highly variable in ground color, ranging from green to pink, and have mottled markings on each segment (Figure 39). The larvae can be found in our area by searching the flower heads of Waxflower (also called Cliffbush) for signs of feeding damage. This species overwinters in the pupal stage.

56. *Glaucopsyche lygdamus oro* Scudder — Plate IV, Fig. 56a-c
SILVERY BLUE

The silvery blue males of the Silvery Blue are aptly named. They are among the first butterflies to appear in the spring, flying from late May into early June, while females may be found until the first week of July. It occurs in fair abundance in the wet meadows, flying with the Greenish Blue *(Plebejus saepiolus)* and the High Mountain Blue *(Agriades franklinii)*.

The hostplant in our area is unknown to date, but elsewhere in Colorado it feeds on Loco-weed *(Oxytropis)*, and milk vetch *(Astragalus)*. The mature larva is green in ground color and is covered with short white hairs (Figure 40). It has a distinctive reddish-pink band dorsally bordered on either side with small yellow triangles. The species flies in a single brood. Winter hibernation occurs in the pupal stage.

Some lycaenid larvae, including those of the Silvery Blue, have special dorsal glands that secrete a sweet fluid. This substance attracts and appeases ants. Pierce and Mead (1981) discovered that Silvery Blue caterpillars tended by ants had a higher survival rate than those from which ants were excluded; these ants protect the larvae from parasitic insects such as wasps.

57. *Lycaeides melissa melissa* (W. H. Edwards) — Plate IV, Fig. 57a-c
ORANGE-MARGINED BLUE

Orange-Margined Blue males are bright, silvery, lilac blue on the dorsal surface and white ventrally, with a well-developed band of orange spots on both the forewing and hindwing ventral surfaces. These bands are even more developed in the female, where the ground color is usually brown on the dorsal surface and the orange spots occur both dorsally and ventrally. This species flies from mid-July to mid-August; the males emerge about five days before the females. The males are much more abundant than the females (often a 5:1 ratio), which may be an artifact due to the more retiring habits of the females. The Orange-Margined Blue is most often encountered in dry meadow habitats, but it also flies in wet meadows.

The larval foodplants include various species in the family Fabaceae, including loco-weeds *(Oxytropis)*, milk vetches *(Astragalus)*, vetches *(Vicia)*, and lupines *(Lupinus)*.

The eggs are laid on the leaves and diapause occurs in the third or fourth larval instar. The mature larva is highly variable in color, usually green in ground color with red and white markings, and the body is covered by fine white hairs. The pupa may be either a plain light green or brownish, depending on the site of pupation. A silk button and girdle hold the pupa in place.

The Orange-Margined Blue flies from mid-June through the third week of September in the Florissant area, with a population peak between mid-July and mid-August. The extended emergence period suggests that there are two broods. One adult has been noted as a prey item of a robber fly (Asilidae) in the Monument area (M. Minno, pers. obs.).

FIGURE 41. The Orange-Margined Blue (*Lycaeides melissa*), a common lycaenid butterfly of dry forest habitats. It frequently stops to take nectar from composite flowers such as this fleabane.

58. ***Plebejus saepiolus whitmeri*** **Brown** Plate IV, Fig. 58a-c
GREENISH BLUE

The Greenish Blue is distinguished by its silvery violet-blue males having a slight greenish tinge on the dorsal surface. The females are generally dark in color but with a considerable area of blue on the dorsal surface, characteristic of this subspecies. This is a very abundant lycaenid in the wet meadow areas of the Florissant region. The first brood flies in June and early July, the second at the end of July and early August. Specimens of the second brood are slightly larger than those of the first brood. The flight is fairly rapid but close to the ground. Adults nectar frequently on clover and other flowers.

Larvae feed exclusively on various species of wild clover *(Trifolium)*. Females deposit the eggs singly on clover flowers. No detailed description of the life history has been published, although there are known to be both red and green color phases in the California larva of this species (Emmel, Emmel, and Mattoon, in press), and diapause apparently occurs as a half-grown larva.

59. ***Agriades franklinii rustica*** **(W. H. Edwards)** Plate IV, Fig. 59a-c
HIGH MOUNTAIN BLUE

The High Mountain Blue can be distinguished from the Greenish Blue *(Plebejus saepiolus)* by its shiny gray-blue upperside and wide, diffuse, dark-scaled outer border of the forewing. The female is entirely brown on the dorsal surface. The undersides of both sexes are highly variable in maculation, with white bands, white spots, or black pupiled spots scattered across the hindwing surface. This species is abundant in wet meadow habitats. Males and females are present from mid-June to the end of July, and fresh specimens may be collected throughout this period. There is only one generation per year, with an extended emergence period.

The only known larval foodplant is Rock-jasmine (*Androsace septentrionalis*) (Primulaceae), which is also used by the same blue species in Europe. The early stages of our subspecies are undescribed in the literature.

Family RIODINIDAE
THE METALMARKS

The Riodinidae are considered by some authorities to be a subfamily of the Lycaenidae, to which they are obviously closely related. Here we treat them as a separate family, which numbers about 1,500 species and is worldwide in distribution but occurs mostly in the American tropics. There are only 20 North American species, primarily from the southern states, and only one species occurs in our area. Riodinids may be distinguished by the fact that the male forelegs are less than half the length of any of the remaining four legs. Most species are extremely colorful and the tropical fauna is remarkably diverse in pattern and form. Metalmarks are swift and erratic flyers but rarely fly far before stopping again. They visit flowers for nectar and usually rest, bask, and feed with their wings spread flat open. Larvae are similar to those of the Lycaenidae, but are less sluglike. Hostplants include a great variety of dicotyledons.

60. ***Apodemia mormo duryi*** **(W. H. Edwards)** Plate IV, Fig. 60a-c
MORMON METALMARK

The Mormon Metalmark occurs as a great number of subspecies and local population forms across the western United States from the Great Plains to coastal California. Its principal habitats lie between 6,000' and 8,000' elevation. Our few records from the

Florissant region are apparently strays from the Dome Rock colony or the next nearest known colony immediately south of Cripple Creek at an elevation of about 8,800'. Specimens from these populations are referable to subspecies *duryi* which also occurs in New Mexico. The dorsal hindwing surface has a distinctive red-orange or rust-colored band bordering the white discal spots. The overall appearance of the butterfly cannot be confused with any other in our area. Adults are almost always found in close association with their larval foodplants, which are various wild buckwheat *(Eriogonum)*. They avidly feed on nectar from these flowers, as well. The butterfly is single-brooded, occurring from late July until September.

In southern Colorado, the generally distributed larval foodplant of this Upper Sonoran Zone butterfly is wild buckwheat (*Eriogonum jamesii*). The pinkish eggs are laid in the flowers and the larvae feed on the flowerheads. The mature larva is purple to dark violet in ground color, somewhat lighter below the spiracles. It has four subdorsal rows of long black tufted hairs and a lateral row of even longer hairs. The butterfly diapauses in the pupal stage. Adults have a very rapid flight but usually do not fly too far when disturbed from their perching sites on wild buckwheats.

Family NYMPHALIDAE
THE BRUSH-FOOTED BUTTERFLIES

This huge family is almost too large and diverse to support generalizations. Numbering over 6, 640 species from around the world (Shields 1989b), nymphalids range from very small to very large in wing span, although most of the 185 North American species are of medium size. There are nine subfamilies, of which five occur in North America and four in the Florissant region: the Heliconiinae (Heliconians or Long-wing Butterflies) with one species, the Nymphalinae (Spiny Brush-footed Butterflies) with 25 species, the Satyrinae (Satyrs and Wood-nymphs) with 9 species, and the Danainae (Milkweed Butterflies) with 2 species. Because of the diversity of adult structural differences and life histories in this group, many authorities elevate these subfamilies to family rank.

The name "brush-footed" comes from the fact that the forelegs in both sexes are very small and are covered with hairs, resembling tiny brushes. Adults walk only on the four mid and hind legs. Most species are strong fliers and many migrate, including the famous Monarch, which completes a yearly round-trip migration from Canada to Mexico and back over several generations. The majority of nymphalids rest and bask with wings spread open, although the satyrines usually keep their wings closed and often bask sideways to the sun.

Larvae vary considerably in shape and coloration, but most of our species bear rows of spines or hairs. The range of foodplants eaten is great, although there is considerable specialization at the tribal and generic levels. Satyrine larvae feed mostly on monocots, but the remaining subfamilies primarily use dicotyledonous plants. The Danainae, as their common name implies, feed on milkweeds, from which they sequester poisonous compounds to use in their own defense as protection against potential predators.

Subfamily HELICONIINAE

61. ***Agraulis vanillae incarnata*** **(Riley)** Plate V, Fig. 61a-c
GULF FRITILLARY

The Gulf Fritillary is typically a tropical and subtropical species that enters Colorado as a stray or a migrant. Occasionally during the summer, this silver-spangled beauty is found flying across our dry meadows. We have records from late June to the end of July, with as many as three somewhat worn specimens taken in one day.

The foodplant of this butterfly is passionflower (*Passiflora*) (Passifloraceae), which does not grow near Florissant because of the cold winters. The larvae are brownish-yellow in ground color and have brown stripes as well as six rows of branching spines. This species is not capable of overwintering in Colorado in any stage, although adults can survive temperatures close to freezing in the southern states.

Subfamily NYMPHALINAE

62. ***Polygonia interrogationis*** **(Fabricius)** Plate V, Fig. 62a-c
QUESTION MARK

The Question Mark is the largest of our anglewing species and is easily identified by the broken silver mark that forms a stylized question mark on the ventral surface of the hindwing. It is rarely encountered in the Pikes Peak area, apparently entering as a stray from lower elevations. Its range is recorded from southwestern New Mexico north through Colorado to eastern Wyoming in the Rocky Mountains. Three specimens taken on 24 June 1987 just west of High Trails Ranch ("Top-of-the-World," at 9,200') by T. Emmel and J. L. Nation Jr, and a single male taken just west of Big Spring Ranch at 9,200' by B. Drummond on 25 June 1987, represent the first records for Park County in Colorado.

The adults hibernate through the winter and appear on the first warm days in spring. They lay eggs on several larval foodplant species, including nettles *(Urtica)* (Urticaceae), that are available at our elevations, but apparently there is no breeding colony. The larvae are reddish-brown in ground color, are dotted with irregular lighter markings, and have branching spines arranged in seven rows. Several broods occur each year at lower elevations. Adults from the summer brood are darker dorsally and more strongly patterned ventrally than those of the "winter" brood (which emerge in the fall, hibernate in winter, and fly again in the spring). Like other anglewings, the adults visit sap flows at wounds in the bark of deciduous trees, or sip fluids from rotting fruit and animal feces.

63. ***Polygonia satyrus* (W. H. Edwards)** Plate VI, Fig. 63a-c
SATYR ANGLEWING

The Satyr Anglewing is distinguished by its golden brown color above, with warm brown markings on the ventral surface. The silver mark on the underside of the hindwing tends to be C-shaped. This species occurs in several color forms in the same population, with adults ranging from those with strongly patterned undersides to ones with pale tan or uniformly golden yellow undersides. It is an uncommon butterfly in the Pikes Peak region and is probably at the upper end of its distributional range from the riparian canyons near Colorado Springs where it is more abundant.

The larvae feed on nettle *(Urtica)*. The eggs are pale green and are deposited singly or in small strings on the underside of the nettle leaves. The mature larva is blackish-brown in ground color, has seven rows of spines, and bears a greenish-white dorsal stripe. There are two short horns on the head. The larva forms a nest out of leaves by using silk threads. The pupa is brown with a few metallic gold spots on the back. As is typical of the nymphalids, the pupa is attached to the substrate only at the tip of the abdomen, and hangs in a head-downward position. The Satyr Anglewing hibernates in the adult stage and adults may be seen on warm days in late spring. This species may be taken occasionally throughout the summer.

64. ***Polygonia faunus hylas* (W. H. Edwards)** Plate VI, Fig. 64a-c
FAUNUS ANGLEWING

The Faunus Anglewing has bright mossy green spots on the dark gray-brown ground color of the ventral surface. The silver mark on the underside of the hindwings is L-shaped. This subspecies is found from central Colorado south into New Mexico and Arizona and north into southern Wyoming; *hylas* was formerly considered to be a separate species by many authors. Early in this century, many specimens of another, darker subspecies, *P. faunus rusticus* (W. H. Edwards), were taken in Colorado, but museum collections have no records for this subspecies for the last quarter century (Ferris and Brown 1981). The *hylas* subspecies has been taken by us at irregular intervals in the Florissant region since 1960.

As with the other *Polygonia* species, the adults hibernate and fly on the first warm days in the spring. They are solitary in the forest environments, and are usually found in the Florissant region only in the cooler canyons below rocky bluffs, where willows, alders, Douglas-fir, and Colorado Blue Spruce grow.

The larvae are known to feed on willows *(Salix)*, alders *(Alnus)* (Betulaceae) and currants *(Ribes)* (Grossulariaceae). The mature larva is reddish or yellowish brown in ground color, with a white saddle over the first two abdominal segments and a dashed dull orange lateral band. The seven rows of spines are whitish at the base and brown near the tips.

65. *Polygonia zephyrus* (W. H. Edwards) — Plate VI, Fig. 65a-c
ZEPHYR ANGLEWING

The Zephyr Anglewing is a common butterfly in our region, flying from the end of snowmelt in May or early June to as late as August or early September (Figure 42). The dorsal color is bright orange marked with brownish to blackish spots and bands, while the underside is a mottled gray. The silver mark on the underside of the hindwing is L-shaped.

This butterfly overwinters in the adult stage. It lays its eggs on currants (*Ribes*). The mature larva is black with reddish-buff markings near the head and whitish markings down the back. Seven rows of spines are present (Figure 43).

This is one of the earliest butterflies of the Rocky Mountain spring, with adults flying on the first few warm days. A mass emergence of adults (all in fresh condition) takes place in the last week of July, after the larvae from the eggs laid by the overwintering adults reach maturity. In the early part of the summer, the adults frequently visit flowers of Shrubby Cinquefoil *(Pentophylloides floribunda)* and occasionally thistles (*Cirsium* species) in dry and wet meadows. Later in the summer, they frequent blooming plants of Rabbit Brush *(Chrysothamnus nauseosus)*.

66. *Nymphalis (Nymphalis) antiopa antiopa* (Linnaeus) — Plate VII, Fig. 66a-b
MOURNING CLOAK

The Mourning Cloak is an inhabitant of streamside canyons in our area, flying along streams and in among willow thickets. It is easily identified by its maroon-black ground color and pale yellow margins with an interior row of bright blue spots. It is a large butterfly and has a distinctive gliding flight alternating with strong wingbeats.

The adults overwinter and become active on the first warm days of spring. The eggs are laid in clusters on willows *(Salix)* and Quaking Aspen *(Populus tremuloides)*. The mature larva is jet black with seven rows of branching spines and a mid-dorsal row of red spots with lateral dots on the side. The pupae are gray-brown, and also bear spines. Adults fly throughout the summer. They readily suck juices from sap flows and rotting fruit, and occasionally visit flowers for nectar.

67. *Nymphalis (Aglais) milberti milberti* Godart — Plate VII, Fig. 67a-b
MILBERT'S TORTOISESHELL

Milbert's Tortoiseshell is immediately recognizable by its broad dorsal yellow and orange band on a rich dark brown background. It occurs across much of North America. In Colorado, adults fly throughout the summer, and range across all dry and wet meadow habitats, with occasional specimens being encountered in forest glades. In contrast, in July the larvae are found only along the banks of semi-permanent streams where the larval foodplants (nettles) occur.

FIGURE 42. The Zephyr Anglewing *(Polygonia zephyrus)* is a common inhabitant of the Ponderosa Pine forest and adjacent meadow habitats in our area.

FIGURE 43. Last instar larva of *Polygonia zephyrus* (Species 65) feeding on *Ribes* in Teller County, Colorado.

The local foodplant is Stinging Nettle, *Urtica dioica* ssp. *gracilis*. The eggs are laid in clusters on the leaves of the host, and the larvae are gregarious in the younger stages. The mature larva is black with greenish-yellow lateral stripes, whitish dots, and black spines (Figure 44). The pupa is brown with a few golden spots on the back. Hibernation occurs in the adult stage. Sometimes these butterflies fly even in mid-winter on warm and sunny days.

FIGURE 44. Fifth instar larva of *Nymphalis milberti milberti* (Species 67), in Teller County, Colorado, on *Urtica dioica* ssp. *gracilis*.

68. *Vanessa virginiensis* (Drury) AMERICAN PAINTED LADY

Plate V, Fig. 68a-c

The American Painted Lady or Virginia Lady has two greatly enlarged hindwing spots on the ventral surface and a rich white spiderweb-like network of lines on the ventral hindwing surface. Solitary adults are found throughout the summer from June to early September.

The larval foodplants include many species in several families, but the everlastings in the family Asteraceae (such as *Antennaria* and *Gnaphalium*) are preferred. Eggs are laid singly on the leaves and the larvae live singly in silken nests, usually at the top of the foodplant in the cluster of flowers or buds. The mature larva is velvety black with blackish spines and narrow yellow crossbands across each segment (Figure 45). There are also two rows of white spots down the back. The head is black. Pupation often occurs within the larval nest. The pupa may be either gray or iridescent greenish-yellow with brown markings (Figure 46). Hibernation occurs in the adult or possibly the pupal stage, depending on the brood.

FIGURE 45. *Vanessa virginiensis* (Species 68), lateral view of last-instar larva feeding on *Gnaphalium pennsylvanicum* (Broward County, Florida).

FIGURE 46. *Vanessa virginiensis* (Species 68), lateral view of pupa (from Broward County, Florida).

69. *Vanessa cardui* (Linnaeus) PAINTED LADY

Plate V, Fig. 69a-c

The Painted Lady is one of the most widely distributed butterflies in North America, being found from Mexico north to Hudson Bay in Canada. It is also found virtually everywhere else in the world as either a resident or a migrant. In Colorado, this species is reintroduced annually through migration and has been recorded from every county. It breeds in large numbers in Mexico, and flies north in the spring, with those adults starting several generations during the summer, followed by a return migration (Emmel and Wobus 1966) in September or October. During the summer months, this species is commonly found in the wet and dry meadows of the Florissant region, visiting thistle blooms.

Thistles (*Cirsium*) (Asteraceae) are the preferred larval foodplants. The eggs are laid singly and the larvae live singly in nests constructed from silk and leaves. The mature larva is greenish-yellow, mottled with black, with several rows of yellowish spines and a lateral yellow stripe. The head is reddish. Pupae are gray with gold-colored bumps. At least two broods occur in our area, with fresh adults being taken from June to August.

70. *Vanessa atalanta rubria* (Fruhstorfer) RED ADMIRAL

Plate V, Fig. 70c-c

The Red Admiral butterfly is encountered throughout the summer and is distinguished by its red-orange bands across the black forewings and the orange-red marginal hindwing band. It occurs across North America and into North Africa, Europe, and western Asia. In our area, it is usually seen in dry meadows where there are many blooming thistle plants *(Cirsium)*, its favorite nectar source.

The larval foodplants are nettles *(Urtica)*. Mature larvae are black, studded with rows of hundreds of whitish velvety points and seven rows of pale spines. The larva ties the edges of a leaf together with silk for shelter and a pupation site. The pupa is grayish brown with gold bumps on the back. Both the adults and pupae may hibernate through the winter. The flight is strong and rapid, but the species may be easily observed or netted while visiting flowers or rotten fruit (at lower-elevation sites). Males are known to be territorial in the late afternoon, defending small areas of sunlit ground.

71. *Euptoieta claudia* (Cramer) VARIEGATED FRITILLARY

Plate V, Fig. 71a-c

The subtropical and tropical Variegated Fritillary is perhaps a rather surprising discovery for the Colorado mountains, but it occurs fairly commonly in the Pikes Peak area throughout the summer. Elsewhere, it ranges from the New England states south into the tropics and west to the Pacific Coast. The Variegated Fritillary is a medium-sized, brownish butterfly. Although it resembles fritillary species in the genus *Speyeria*, it lacks their characteristic white or silver spots on the underside of the hindwings.

The foodplants include violets *(Viola)* (Violaceae) and Stone Crop *(Sedum lanceolatum)*, and in our area it is frequently seen ovipositing on Wild Blue Flax *(Adenolinum lewisii)* (Linaceae). The larvae are bright orange-red with two dark lateral stripes on each side, and with white spots. There are six rows of dark bluish-black branching spines, with a lengthened pair protruding forward from the thorax over the head. The pupa is white, with orange and black markings.

Several broods occur in the southwest, and Florissant region specimens have been recorded from the 2nd of June to the 15th of September. This species apparently passes the winter in the adult stage, although it is not known whether it is able to overwinter in Colorado. In the larger dry meadows of our area, *Euptoieta claudia* flies rapidly from thistle to thistle, sipping nectar.

72. ***Speyeria idalia*** **(Drury)** — Plate VII, Fig. 72a-c
REGAL FRITILLARY

The Regal Fritillary formerly inhabited wet meadow areas across the northeastern United States and a few tall-grass prairie regions in eastern Colorado. Single specimens of this species have been taken at several sites on the plains in Boulder, Larimer, and El Paso counties, as well as extreme northeastern Colorado (Ferris and Brown 1981). On August 2, 1987, T. C. Emmel and J. L. Nation, Jr. discovered a fresh female *S. idalia* flying across a ridgetop meadow at 8,710' elevation near Fish Creek on Big Spring Ranch, south of Lake George, in Park County. The flight was slow and stately, almost like a Monarch. The immediate habitat in this area is a relatively dry meadow and a gentle mountain slope; however, a tall-grass, wet-meadow habitat occurs within a short distance and violets (*Viola* species), the known larval foodplants, occur here. Four other *Speyeria* species feed on the violets on the ranch in these wet meadow habitats. The freshly emerged condition of the *idalia* female taken in 1987 suggests a local emergence.

The larva of this species is dark velvety brown in ground color, with six rows of barbed blackish spines and yellow spots. The pupa is brownish with darker markings, and it hangs from the cremaster.

73. ***Speyeria edwardsii*** **(Reakirt)** — Plate VII, Fig. 73a-c
EDWARDS' FRITILLARY

Edwards' Fritillary is about as abundant as the Atlantis Fritillary *(Speyeria atlantis)* in our region. In contrast to that species, it prefers forest margins. The females of *edwardsii* habitually fly along the forest edge among the trees, while the males can be found sipping thistle nectar between the edge of the forest and perhaps 100-200 yards out from the last trees. This species has a stately, strong flight and is not easily captured except while feeding. It is a large and tawny fritillary with bold black borders on the dorsal surface, especially in the females. The greenish ventral surface of the hindwing bears large, elongate silver spots and is also quite distinctive in pattern from the other fritillary species.

As in other *Speyeria*, the larval foodplants are violets. In our area, the larvae feed on Mountain Blue Violet *(Viola adunca)*, but at lower elevations *Viola nuttallii* is the foodplant (Scott 1986). Eggs are laid singly on or near violets. Mature larvae are patterned with yellow, gray, and black, and have long branching spines. The first instar larvae hatch from the eggs, then go into diapause without feeding until the following spring.

74. *Speyeria coronis halcyone* (W. H. Edwards) Plate VII, 74a-c
CORONIS FRITILLARY

The Coronis Fritillary is a large species with tawny ground color and a rich dark red-brown ventral coloration in the disc area of the hindwing. The silver spots are quite large and elongate, as in Edwards' Fritillary. The tawny ground color of the dorsal surface and the dark red-brown ventral surface of the hindwing will distinguish it, however.

This species characteristically prefers hillsides covered with shrubby growth in the transition zone, at slightly lower elevations (7,500' or below) than are found in the Florissant region. Thus, our specimens, which occur occasionally in July and early August, are probably strays from lower elevations near Colorado Springs. The adults have a swift flight and are very difficult to catch unless they stop to feed at flowers such as thistles. The larvae feed on violets.

75. *Speyeria atlantis electa* (W. H. Edwards) and *hesperis* (W. H. Edwards) Plate VII, Fig. 75a-d
ATLANTIS FRITILLARY

The Atlantis Fritillary occurs along the Front Range in Colorado in two color forms. Both the *electa* and *hesperis* forms have the same type locality (Turkey Creek Junction, Jefferson County). The subspecies form known as *electa* has a dark discal area with bright silver spots on the ventral hindwing surface. The *hesperis* phenotype has a medium-brown disc with opaque (unsilvered) spots on the ventral hindwing surface. Both color forms occur on Big Spring Ranch. This species prefers the open dry meadow areas and adults are often found feeding on white thistles. Females search among the aspens in wet meadows for violets (*Viola*) on which to lay their eggs. Both males and females appear from the first of July to the second week of August, and breeding colony areas are highly restricted in the relatively dry Pikes Peak region.

Like the other *Speyeria* species, the larva of *S. atlantis* is blackish in ground color, with six rows of dark spines and yellow spots.

76. ***Speyeria mormonia eurynome* (W. H. Edwards)** Plate VII, Fig. 76a-c
MORMON FRITILLARY

This is the smallest of the five *Speyeria* species in the Florissant region. It flies from mid-July to late August, and, as far as is known, it is restricted to breeding in one small meadow area (elevation 8,600') about one-half mile north of the Big Spring Ranch headquarters. This species is much more abundant in the elevations above 9,000' elsewhere in Colorado.

The larvae are black with yellow spots and six rows of branching spines, and feed on violets.

77. ***Phyciodes tharos pascoensis* Wright** Plate VI, Fig. 77a-c
PEARLY CRESCENTSPOT

The Pearly Crescentspot lacks the pale transverse bar found on the ventral forewing surface of the Field Crescentspot *(P. campestris)*. This butterfly ranges over most of North America. In Colorado, it occurs from riparian canyons near Colorado Springs to high in the mountains. Adults fly in wet meadows during June and July, but the species is never common in the Florissant region.

The eggs are deposited in clusters on leaves of *Aster* species (Asteraceae). In the young larvae are gregarious and overwinter together at the end of the third instar. Mature larvae are black with yellow dots, and a yellow lateral band. They are covered with spiny yellow-brown tubercles. The pupa is light grayish-brown.

78. ***Phyciodes campestris camillus* Edwards** Plate VI, Fig. 78a-c
FIELD CRESCENTSPOT

The Field Crescentspot has a more contrasting pattern than the Pearly Crescentspot. It has a wide range throughout the western United States, from sea level to above treeline. It is normally taken in wet meadow habitats or on flowers in adjacent hillside meadows.

The larvae feed on asters (*Aster*) and the eggs are deposited in clusters. The dark larvae diapause at the end of the third instar and complete development the following spring. There is apparently only one brood at our elevation, flying in June and very early July.

79. ***Phyciodes vesta* (W. H. Edwards)** Plate VI, Fig. 79a-c
VESTA CRESCENTSPOT

This species is typically a lowland Texas butterfly, where it is associated with riparian canyons and river bottoms. Migrants occasionally fly into the mountain area of Colorado and are found in the Pikes Peak-Florissant region. All our records are for the month of July.

The life history of the Vesta Crescent is not formally described, but the larval foodplant in Texas is known to be *Siphonoglossa pilosella* (Acanthaceae) (Scott 1986), a plant that does not occur in Colorado. The adults of this occasional migrant are easily recognized by the finely checkered dorsal markings and the banded ventral pattern on both wings.

80. ***Phyciodes pallidus* (W. H. Edwards)** Plate IX, Fig. 80a-b
PALE CRESCENTSPOT

The Pale Crescentspot is somewhat larger than our other local crescentspots. The wings are lightly marked with black on the upperside and have silvery white submarginal spots on the underside. Its favored habitat is slightly lower than our area, in the foothill and Transition Zone valleys. It was first captured in 1989 in the Florissant area in two places: Maze Caves (June 26, by Andrew Warren) and Big Spring Ranch (July 27, by T. C. Emmel).

FIGURE 47. The Pale Crescentspot, *Phyciodes pallidus* (W. H. Edwards): adult female, dorsal (left specimen) and ventral surfaces, captured in 1989 at Maze Caves on Big Spring Ranch near the western boundary of Florissant Fossil Beds National Monument.

Elsewhere in Colorado, this butterfly is single brooded and flies from late May through June. The larvae, ochre with a brown middorsal line and a brown band above the spiracles, feed on thistles *(Cirsium)* in the Asteraceae.

81. *Chlosyne (Charidryas) gorgone carlota* (Reakirt) Plate VI, Fig. 81a-c
GORGONE CHECKERSPOT

The Gorgone Checkerspot has a unique zig-zag pattern across the outer half of the ventral surface of the hindwing and has more rounded wings than species of *Poladryas* or *Euphydryas*. It is basically a lowland butterfly that is found only occasionally up to 10,000' in Colorado. Specimens in the Florissant area are rare and it is uncertain as to whether it is a breeding resident.

Our records are from early to mid-June, in wet meadow habitats along streams. Recorded foodplants elsewhere are asters (*Aster*) and several sunflowers in the genus *Helianthus* (Asteraceae), both of which occur in our area, although these rarely develop to the blooming stage before August. Eggs are laid in clusters. Hibernation occurs in the third larval instar. The mature larva is yellowish in ground color with three longitudinal black stripes and black barbed spines across the body surface.

82. *Chlosyne (Charidryas) palla calydon* (Holland) Plate VI, Fig. 82a-c
PALLA CHECKERSPOT

The Palla Checkerspot has a typical checkerspot pattern of orange- or reddish-brown and black on the upperside of the wings. The hindwings are whitish with red bands beneath. This species is found in low numbers in wet or semi-dry meadows from late June to late July.

The larvae have five rows of black branching spines similar to those of other checkerspots, and are blackish in ground color. They are reported to feed on the leaves of *Aster* species at other sites in Colorado. Hibernation occurs in the third larval instar. The pupa is light brown with pale brown mottling.

83. *Chlosyne (Thessalia) fulvia* (W. H. Edwards) Plate VI, Fig. 83a-c
FULVIA CHECKERSPOT

Taxonomic placement of the Fulvia Checkerspot is controversial. *Chlosyne alma* and *C. fulvia* have been treated as different species, or as subspecies of *C. leanira*. Ferris and Brown (1981) follow Higgins (1960) and place *alma* and *fulvia* as subspecies of *leanira*. In that case, our subspecies would be *C. l. fulvia,* a butterfly that occurs in southern Colorado and northern New Mexico as well as the type locality of Texas. Until a thorough revision of this checkerspot group is done, including larval stages, hostplants, and genetic relationships, the name of our butterfly will be uncertain.

This butterfly is orangish dorsally, with considerable black overscaling in some specimens. The underside is more streaklike than checkered, with black, orange and elongate white markings between the dark veins. This species is readily recognized on the underside by the black veins, orangish forewing and whitish hindwing and off-white background color on the hindwings. The overall appearance is of longitudinal white bands or long blocks of color rather than the small checks of the other checkerspots in our area. The Fulvia Checkerspot is a breeding resident of the drier, lower slopes just south of Cripple Creek and strays into our area only on occasion. Specimens have been recorded in July, usually in the last two weeks.

The foodplant for *C. fulvia* in Colorado is Orange Paintbrush (*Castilleja integra*) (Scrophulariaceae). The larva is ochre in ground color, bearing black spines and black patches at the bases of the spines. The winter is passed in diapause at the end of the third instar. The pupa is white with black marks and spots. This butterfly is known to have breeding populations up to 9,000' in Colorado, and it may have one or more as-yet-undiscovered breeding populations in the Florissant region.

84. ***Poladryas arachne arachne* (W. H. Edwards)** Plate VI, Fig. 84a-c
ARACHNE CHECKERSPOT

The Arachne Checkerspot is characterized by a tawny dorsal color with black markings. Underneath, the ventral surface of the hindwing bears a pearly white band with curved black markings. It is a distinctive checkerspot that occurs in numbers throughout the Florissant region from mid-July through mid-August, particularly in the dry meadow habitat. Male and female flight periods overlap almost completely, although the males begin flying slightly earlier than the females. There is a fair degree of seasonal variation in size, with smaller specimens being found in July and larger males and females in August.

The larvae feed on beard-tongue *(Penstemon)* (Scrophulariaceae), and are whitish in ground color with black spines (Sperry and Sperry 1932). The pupa is white, marked with black blotches (Comstock 1958). The species overwinters in the half-grown larval stage. Apparently, the species is univoltine in the Florissant area, although it may be bivoltine at lower elevations in the Rocky Mountains.

85. ***Euphydryas (Occidryas) anicia capella* (Barnes)** Plate VI, Fig. 85a-c
ANICIA CHECKERSPOT

This abundant checkerspot butterfly is represented by at least two basic phenotypes in our area, which may represent separate species with further study. This butterfly reaches its peak of abundance in mid-July. At that time, it may be found virtually everywhere in this region: valley and montane dry meadows, wet meadows, and open forests. Males begin emerging a week or more before the females. Females reach their peak abundance in late July.

In the wet meadow habitats of the Florissant area, the adults feed on nectar from the yellow flowers of Shrubby Cinquefoil, *Pentaphylloides floribunda*. In the dry

FIGURE 48. An adult male *Euphydryas anicia capella*, resting momentarily on a black-eyed susan.

meadow and forest areas, the Anicia Checkerspot commonly visits yellow blooms of butterweed (several *Senecio* species). The flight is not particularly strong, and large numbers may be collected with ease.

As noted above, the variation in populations in the Florissant area is considerable. The earliest specimens in June are strongly reddish in tone, resembling *E. a. capella* from lower elevations (type locality: Manitou Springs), while later fliers in July have more yellow on their wings. Variation in wingspread is also considerable. A number of specimens with aberrant wing patterns, ranging from practically all red to practically all black, have been found each summer.

Curiously, despite the species' abundance, the foodplant of this checkerspot butterfly is not known in the Florissant area, although the mature larva and pupa have been described (Emmel 1963) from material collected on Big Spring Ranch. The mature larva is ivory white, mottled with black, and has black bristly tubercles. The larval foodplants elsewhere in the range of *E. anicia* include beard-tongue *(Penstemon)*, paintbrush *(Castilleja)*, and *Besseya* (all Scrophulariaceae). The half-grown larvae diapause through the winter under rocks.

86. *Limenitis (Basilarchia) weidemeyerii weidemeyerii* W. H. Edwards
WEIDEMEYER'S ADMIRAL

Plate VII, Fig. 86a-b

Weidemeyer's Admiral is distinguished by its broad white bands set against a blue-black ground color. It flies primarily in July and is found only around aspen and willow thickets along watercourses. It rapidly returns to this "flyway" if frightened away by a bird or lepidopterist. Both aspen and willow are known larval foodplants.

As the distribution of the willow trees follows the small semi-permanent streams around the Florissant region, it may be inferred that the distribution of the butterfly at almost every available preferred habitat is due to the "corridors" of travel provided by the willow-lined streams. Without these corridors, it is doubtful whether the butterfly could cross large areas of pine forest or dry meadows to become established in new areas (barring, of course, accidental dispersion by wind or other extrinsic factors). Although a strong flyer, this butterfly is restricted by a behavioral barrier from straying out of this particular set of ecological conditions. Occasionally, males are found drinking at mud in wet meadow habitats or at roadside puddles after rainstorms.

The larval hosts include several willows *(Salix)* and Quaking Aspen *(Populus tremuloides)*. The eggs are laid singly on the leaves after the first adults appear in late June and early July. Young larvae are brown with a white saddle and resemble bird droppings. The half-grown larva overwinters in a rolled leaf shelter. The mature larva in the following year has a grayish ground color mottled with lighter patches, and bears a bristly pair of thick tubercles on the top of the second segment behind the head. There is also a distinctive humpback structure on the thorax, which extends into the pupal stage as well (Figure 49).

The adults are quite inquisitive and will swoop down from perches on the ends of willow branches to investigate butterflies and other insects that fly into their apparent territories. This is one of the most distinctive butterflies in the Florissant region and can provide hours of fascinating observation of its behavior.

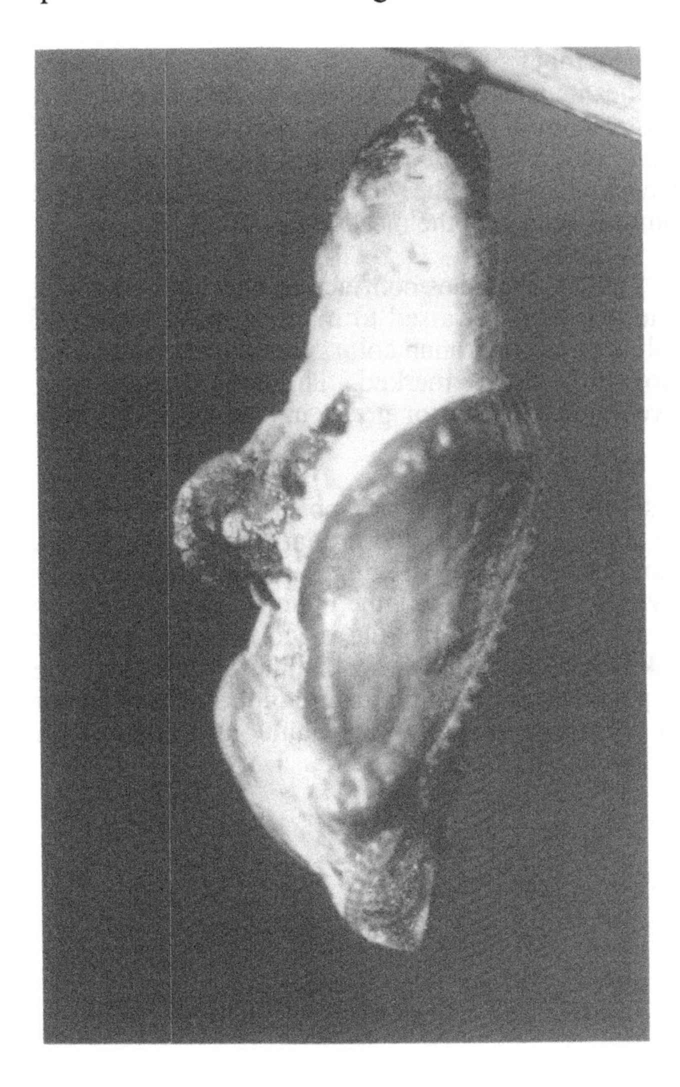

FIGURE 49. *Limenitis weidemeyerii weidemeyerii* (Species 86), lateral view of pupa (Teller County, Colorado).

Subfamily SATYRINAE

87. ***Cyllopsis pertepida dorothea*** **(Nabokov)** Plate VIII, Fig. 87a-c
CANYONLAND SATYR

This species, sometimes known as Dorothy's Satyr, is characteristic of the Lower Sonoran Zone to the south of the Florissant region, below Cripple Creek and toward Canon City, in foothill areas where Gambel's Oak grows. However, strays are found occasionally in the Florissant region during the month of July.

The mature larva is straw-colored, with numerous faint brown lines increasing the resemblance to a dead grass blade. There is a brown sublateral band. The head is straw-colored, with two long hairs having a brown stripe on each side. The pupa is straw-colored, with numerous brown longitudinal striations and two short horns on the head. It feeds on grasses, is single-brooded, and the half-grown larvae overwinter (Scott 1986).

88. ***Coenonympha tullia ochracea*** **W. H. Edwards** Plate VIII, Fig. 88a-c
OCHRE RINGLET

The Ochre Ringlet is an abundant member of our butterfly fauna. This species appears in the first week of June and reaches its peak in the last week of June and the first week of July. At this time it is the most common butterfly in the Florissant region. If desired, more than a hundred individuals could be collected in an hour. The favored habitats of the Ochre Ringlet are the marginal areas of the wet and dry meadows, where scattered aspen and pines extend out from the forest into the meadows.

The larval foodplants are meadow grasses. Females confined in a jar or a cage will lay eggs readily on grasses. The larvae can then be raised to maturity on a variety of grasses. The mature larva varies from olive to tan in ground color, with darker lengthwise stripes. The mid-dorsal stripe is the most prominently marked. The larva has two short tails. The pupa hangs from grass leaves and is brown or green in ground color, with darker stripes on the wingcases.

The adults have a very short proboscis (tongue) and do not appear to feed on flowers, although they frequently land on grass stems and leaves and perhaps sip dewdrops. A frequently observed habit in the morning hours is wing "flexing," in which the male or female repeatedly opens and closes its wings. When startled, the butterfly flies away with a slow but jerky flight. Males patrol all day long across the meadows searching for females. Studies underway at Pikes Peak Research Station show that there is considerable variability in the pattern of ocelli (eye-spots on the wings) in different populations in our area, which may reflect differing natural selection pressures on the adult butterflies or their immature stages.

89. ***Cercyonis meadii meadii*** **(W. H. Edwards)** Plate VIII, Fig. 89a-c
MEAD'S WOOD NYMPH

The red-eyed Mead's Wood Nymph is found primarily along the forest margins of dry meadow areas and flies in a single brood from late July through the end of August. It is slightly larger than *Cercyonis oetus*, especially in the females, and both sexes have a distinctive flush of rusty red on the forewings.

The eggs of this species are laid on grasses and the larvae go into hibernation immediately after hatching from the egg. The mature larva is green with a darker green dorsal stripe and a white lateral stripe on each side. The pupa is dull green in ground color and is usually suspended from several grass blades held together by silken threads.

This species is considerably less common than *C. oetus*, and is more localized than that species, occurring in fairly widely scattered populations in the Florissant area.

90. ***Cercyonis oetus charon*** **(W. H. Edwards)** Plate VIII, Fig. 90a-c
DARK WOOD NYMPH

The Dark Wood Nymph is one of the most abundant butterflies in the Florissant area in late summer. It flies in one brood but adults may start appearing in late June and reach great abundance during late July and early August. Adults are very dark brown on both surfaces, and usually have one or two large eyespots on each dorsal forewing tip. There is a highly variable number of smaller eyespots (ocelli) on the underside of the wings of both males and females, varying from 0 to as many as 6 spots on the hindwings, and from 1 to 4 or even 5 spots on the forewings. These populations have been studied by the senior author (T. C. Emmel) for the last quarter century, and are revealing interesting patterns of microevolutionary change in response to different selection pressures. Adults avidly visit flowers of many species, especially the composites.

Larval foodplants include various grasses. The eggs are laid singly on grasses, at the rate of about 10 per day. The mature larva is green in ground color, with a dark green dorsal stripe narrowly edged with white. The two larval tails are reddish. The pupa may be green or light brown in color, and is heavily striated in the latter case with dark brown stripes. The first-instar larvae hibernate immediately after hatching from the eggs which are laid in the summer. The bouncing sporadic flight of the adults is quite characteristic and these dark butterflies may be seen everywhere across the Florissant region during the last two weeks of July and the first two weeks of August. Males patrol the meadows all day, searching for receptive females with which to mate.

91. *Erebia epipsodea epipsodea* Butler — Plate VIII, Fig. 91a-c
BUTLER'S ALPINE

Butler's Alpine is a characteristic inhabitant of wet meadow habitats throughout the Florissant region. Here, males emerge principally during the last two weeks of June and females emerge chiefly during the last two weeks of July. The males apparently need about 3-4 weeks for their sperm cells to mature before they mate (Emmel 1964). The uniform dark brown dorsal and ventral ground color with reddish patches on both wings are distinctive. The number of ocelli (eyespots) and development of the eyespot at each position are highly variable.

The chalky-white eggs are laid singly on grasses and the larvae feed until hibernation occurs between the third and fourth instars. The mature larva is pale green in ground color, with a darker green mid-dorsal stripe and dark olive-green lateral stripe. There are two short tails at the end of the larval body. The pupa is pale brown in ground color, with spots and blotches of brown or yellow. The low, slow, bouncing flight is highly characteristic of this species.

92. *Neominois ridingsii ridingsii* (W. H. Edwards) — Plate VIII, Fig. 92a-c
RIDINGS' SATYR

Ridings' Satyr is a remarkable satyrine butterfly whose closest relatives are members of the Asian genus *Karanasa*. It has a relatively short flight period in the Florissant area, and its distribution is restricted to the dry meadows at the bottoms of the east and south slopes of Little Blue Mountain or other high places. It is not particularly scarce in its favored habitat during the second week of July. Individuals commonly are flushed from clumps of grass as one walks through the meadow. A flushed individual flies for perhaps 10-15 meters in a low, rapid, erratic flight and then alights on a bare patch of gravel, facing toward the area it just left, and folds its wings tightly over its body. This behavior is strikingly similar to that of *Oeneis* species. Adults are never seen visiting flowers. In fact, the adults rarely fly unless flushed by a human observer. The underside pattern of mottled gray coloration helps them to blend superbly with the environment around them.

The larval foodplants are grasses. The mature larva is olive brown in ground color, with several darker stripes running lengthwise along the body. There are two short tails at the posterior end. The pupa may be green or dull brown and is usually suspended from leaves near the base of the grassy clumps where the larva has been feeding. This species overwinters as an immature larva.

93. *Oeneis chryxus chryxus* (Doubleday and Hewitson) — Plate VIII, Fig. 93a-c
CHRYXUS ARCTIC

For the first several seasons of field work in the Florissant region (Emmel 1964), it was not realized that the populations of the Chryxus Arctic in our area were restricted to flying in even-numbered years (namely 1960, 1962, etc.). Subsequently, it was discovered that *chryxus* populations throughout the region covered by this book are biennial, with adults appearing only every two years. In even-numbered years, this species flies from early June through the first week of July, reaching great abundance in mid and late June. It is found then in every dry meadow, whether in a valley or on a mountaintop. It flies close to the ground, frequently alighting and folding its wings tightly, with its cryptic underside (black and white mottling) camouflaging it to a high degree when it lands on granitic gravel or in dried grasses (Figure 50).

The females oviposit in grasses, laying the white eggs individually. The larva feeds until the end of the second instar, when it goes into hibernation. The mature larva is tan in ground color, with brown lateral and dorsal stripes (Figure 51). The entire larval surface is covered with small dull tan hairs. The stout pupa is tan in ground color with somewhat darker head areas and wing cases. Populations elsewhere in the Rocky Mountains may fly annually or only in alternate years. This species may require two years for development everywhere, but some populations may not be synchronized to a two-year cycle, instead being composed of two temporally separated populations that are out of synchrony.

FIGURE 50. A fresh male Chryxus Arctic (*Oeneis chryxus*) is well camouflaged among the decomposed granitic soils of the dry meadow habitat.

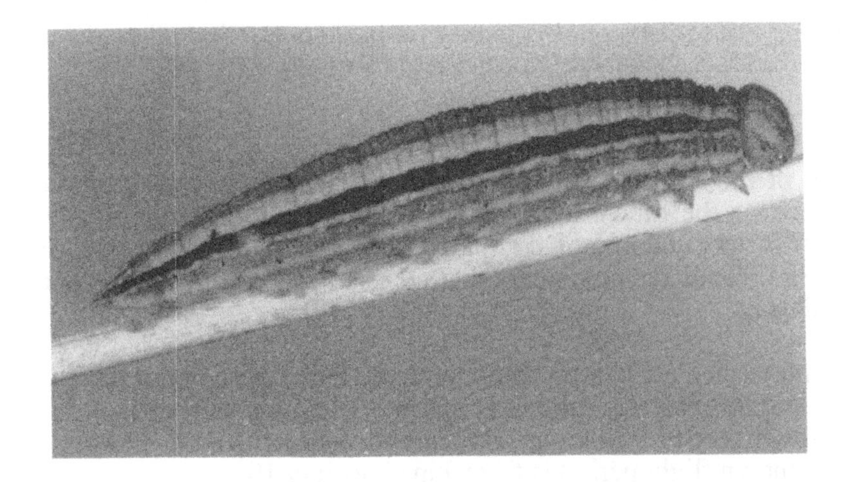

FIGURE 51. *Oeneis chryxus chryxus* (Species 93), lateral view of mature larva (Teller County, Colorado).

94. ***Oeneis uhleri uhleri* (Reakirt)** Plate VIII, Fig. 94a-c
UHLER'S ARCTIC

This subspecies of *O. uhleri* occurs east of the Continental Divide in Colorado and is characterized by the bright tawny color of the dorsal surface and the lack of any well-developed mesial band on the ventral surface of the hindwing (Figure 52). This butterfly may be easily confused with a similarly sized Chryxus Arctic *(O. chryxus),* when in flight,

FIGURE 52. Uhler's Arctic *(Oeneis uhleri)* female resting in grass. Note the very large abdomen, indicating that the female is yet to lay most of her eggs.

and indeed the distribution of the two species is more or less identical in our region. However, the ventral surface of the hindwing is much paler in *O. uhleri* than in *O. chryxus.* Also, *Oeneis uhleri* is very uncommon compared to *O. chryxus* when *chryxus* is flying in even-numbered years. Thus far, it appears that *Oeneis uhleri* is univoltine and appears each year in the adult stage.

The eggs are laid on grasses and are chalky white in color. The larvae begin feeding immediately after hatching and hibernate in the fourth instar. The mature larva is tan in ground color, with dark brown stripes. The pupa is yellowish brown, with a slightly darker abdominal area. The known flight period is from June 2 to July 10.

95. ***Oeneis alberta oslari* Skinner** Plate VIII, Fig. 95a-c
ALBERTA ARCTIC

The Alberta Arctic is a smaller species than the other two *Oeneis* in our area, and is paler and much more gray. The colonies are highly localized in Colorado and our only known population is in Florissant Fossil Beds National Monument in the grassland area. It flies very early in this area, captures being recorded from May 11th to June 11th.

Elsewhere, the life history of this species apparently involves a single annual brood. The eggs are laid on grasses and are grayish-white in color. The mature larva is olive brown in ground color, with darker lateral stripes and one mid-dorsal stripe. Hibernation apparently occurs in the last instar or pupal stage. The pupa is grayish-green in ground color, with darker olive wing cases. Known foodplants are fescues, grasses in the genus *Festuca* (Poaceae) (Ferris and Brown 1981).

Subfamily DANAINAE

96. ***Danaus plexippus plexippus* (Linnaeus)** Plate II, Fig. 96a-b
MONARCH

The Monarch occurs across all of North America during the summer months, although adults in the Rocky Mountain area must return to overwintering colonies in the southern Mexican state of Michoacan. In Colorado, it is an annual summer migrant. In our region, both males and females are found from July to August, soaring high over the dry meadows in stately flight. As many as a half dozen individuals may be observed in one day.

Larvae have been found on milkweed *(Asclepias)* (Asclepidaceae) hostplants that grow along the South Platte River north of Lake George, about seven miles west of Florissant, which is well within the local flight range of this renowned migrant. *Asclepias halli* is an infrequent colonizer of disturbed areas in the Florissant region, including the Pikes Peak Research Station area, but repeated searches of these plants over the past four years have turned up no larvae. The mature larva is quite distinctive, with two pairs of tubercles (one on either end), and alternating greenish-yellow and black bands along the length of the body (Figure 53). The pupa is light blue green with metallic gold dots.

This species is a well-known model species in mimicry complexes with the Queen and Viceroy *(Limenitis archippus* Cramer; Nymphalinae) butterflies. The Monarch larvae that feed on many of the milkweed species obtain poisonous chemicals (cardiac glycosides) from their host plants and pass these along to the adult stage, which renders both larva and adult distasteful to bird predators. Interestingly, some Monarchs feed on non-toxic milkweed species and yet are protected from bird attack by their resemblance to the more common poisonous Monarch individuals.

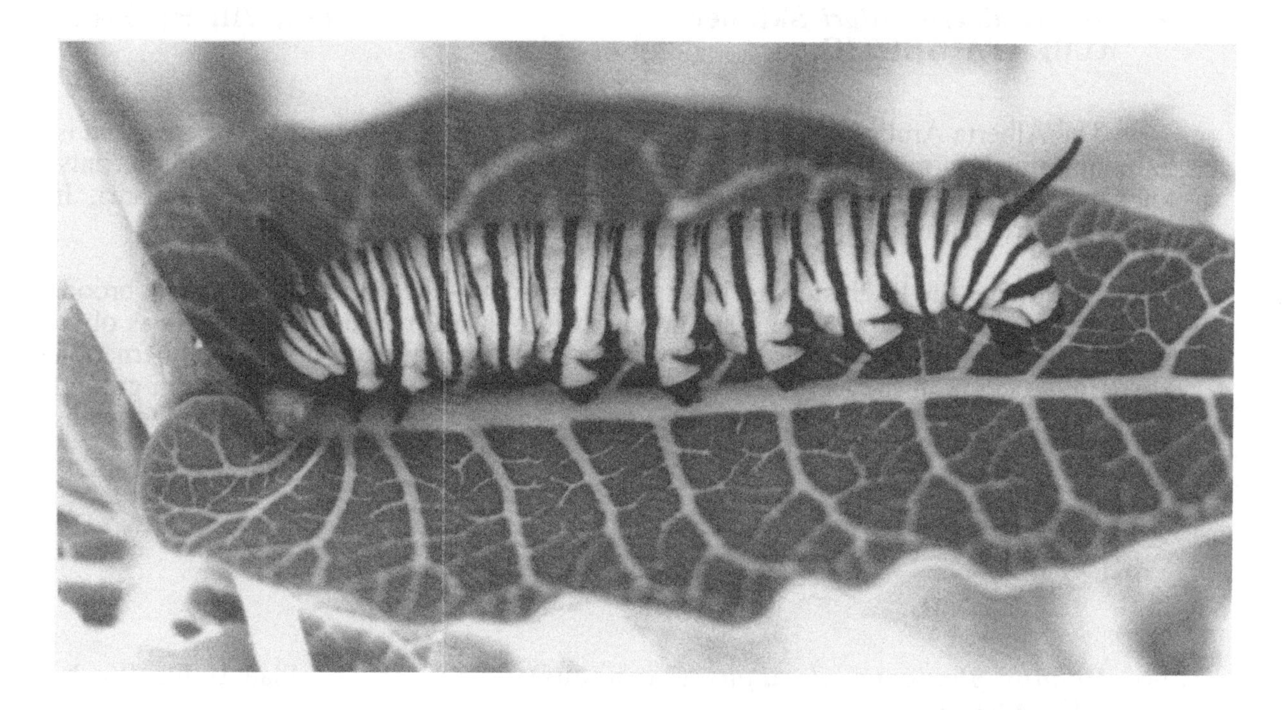

FIGURE 53. *Danaus plexippus plexippus* (Species 96), mature larva from Putnam County, Florida.

97. ***Danaus gilippus strigosus*** **(Bates)** Plate IX, Fig. 97 a-b
QUEEN

The first specimen of the Queen was taken just east of the Nature Place's Sportsplex building on July 6, 1990, by Simon J. Groce, as it was slowly flying across a dry meadow habitat near aspen trees. It was a male in fresh condition.

The species is widely distributed across the southern United States, from the deserts of southern California across Arizona and New Mexico to Texas and the Gulf States. The Queen is not resident as far north as Colorado, but specimens have been taken occasionally on the eastern plains. Like the Monarch, its host plants are milkweeds and related plants (Asclepiadaceae), which are present in the Florissant area. But neither of these tropical danaids can overwinter in a freezing climate.

REFERENCES

Beutenmuller, William, and T. D. A. Cockerell. 1908. *In* Cockerell, T. D. A. 1908. Fossil insects from Florissant, Colorado. *Bulletin of the American Museum of Natural History*, 24: 59-69.

Brown, F. Martin, Donald Eff, and Bernard Rotger. 1957. *Colorado Butterflies*. Denver Museum of Natural History, Denver, Colorado. 368 pp.

Brown, F. Martin. 1976. *Oligodonta florissantensis*, gen. n., sp. nov. (Lepidoptera: Pieridae). *Bulletin of the Allyn Museum,* No. 37: 1-4.

Cockerell, T. D. A. 1907. A fossil butterfly of the genus Chlorippe. *Canadian Entomologist,* 39: 361-363.

Cockerell, T. D. A. 1913. Some fossil insects from Florissant, Colorado. *Proceedings of the United States National Museum,* 44: 341-346.

Cockerell, T. D. A. 1922. A fossil moth from Florissant, Colorado. *American Museum Novitates*, No. 34: 1-2.

Comstock, John A. 1958. A brief note on the pupa of *Melitaea pola* Boisduval (Lepidoptera: Nymphalidae). *Bulletin of the Southern California Academy of Sciences*, 57(3): 143-144.

Durden, Christopher J. 1966. Oligocene lake deposits in central Colorado and a new fossil insect locality. Journal of Paleontology, 40: 215-219.

Durden, Christopher J., and Hugh Rose. 1978. Butterflies from the Middle Eocene: the earliest occurrence of fossil Papilionoidea (Lepidoptera). *Pearce-Sellards Series* (Texas Memorial Museum, Austin), No. 29: 1-25.

Edwards, Mary E., and William A. Weber. 1990. *Plants of Florissant Fossil Beds National Monument.* Pikes Peak Research Station, Bulletin No. 2: iv & 23 pp.

Emmel, John F., Thomas C. Emmel, and Sterling O. Mattoon. In press. *The Butterflies and Skippers of California*. Stanford University Press, Stanford, California.

Emmel, Thomas C. 1963. Notes on the larva and pupa of *Euphydryas eurytion* (Lepidoptera: Nymphalidae). *Bulletin of the Southern California Academy of Sciences*, 62(1): 19-21.

Emmel, Thomas C. 1964. The ecology and distribution of butterflies in a montane community near Florissant, Colorado. *American Midland Naturalist*, 72(2): 358-373.

Emmel, Thomas C., and R. A. Wobus. 1966. A *southward* migration of *Vanessa cardui* in late summer and fall, 1965. *Journal of the Lepidopterists' Society,* 20(2): 123-124.

Ferris, Clifford D., and F. Martin Brown. 1981. *Butterflies of the Rocky Mountain States*. University of Oklahoma Press, Norman. 442 pp.

Gall, Lawrence F., and Bruce H. Tiffney. 1983. A fossil noctuid moth egg from the Late Cretaceous of eastern North America. *Science*, 219: 507-509.

Higgins, L. G. 1960. A revision of the Melitaeine genus *Chlosyne* and allied species (Lepidoptera: Nymphalinae). *Transactions of the Royal Entomological Society of London*, 112(14): 381-475.

Hodges, Ronald W., editor. 1983. *Check List of the Lepidoptera of America North of Mexico*. E. W. Classey and Wedge Entomological Research Foundation, London. 284 pp.

Hovanitz, William. 1950. The biology of *Colias* butterflies. II. Parallel geographic variation of dimorphic color phases in North American species. *Wasmann Journal of Biology*, 8: 197-219.

Kaesler, Roger L. 1987. Superclass Hexapoda. Pages 264-269. *In* Richard S. Boardman, Alan H. Cheetham, and Albert J. Rowell. *Fossil Invertebrates*. Blackwell Scientific Publication, Palo Alto, California. 713 pp.

Leakey, L. S. B. 1953. Minutes of the meeting. Proceedings of the Geological Society of London Session, 1952-1953: 71.

MacGinitie, H. D. 1953. Fossil plants of the Florissant beds, Colorado. Carnegie Inst. Washington, Publ. No. 599, Contrib. Paleontology. Pp. 1-198.

MacNeill, C. Don. 1975. Family Hesperiidae. Pp. 423-578. *In* William H. Howe, editor. *The Butterflies of North America.* Doubleday & Company, New York. 633 pp.

Miller, Jacqueline Y., and F. Martin Brown. 1989. A new Oligocene fossil butterfly, *Vanessa amerindica* (Lepidoptera: Nymphalidae), from the Florissant formation, Colorado. *Bulletin of the Allyn Museum,* No. 126: 1-9.

Miller, Lee D., and F. Martin Brown. 1981. *A Catalogue/Checklist of the Butterflies of America North of Mexico.* Lepidopterists' Society Memoir No. 2. 280 pp.

Pierce, N. E., and P. S. Mead. 1981. Parasitoids as selective agents in the symbiosis between lycaenid butterfly larvae and ants. *Science,* 211: 1185-1187.

Pyle, Robert Michael. 1981. *The Audubon Society Field Guide to North American Butterflies.* Alfred A. Knopf, New York. 916 pp.

Scott, James A. 1986. *The Butterflies of North America.* Stanford University Press, Stanford, California. 583 pages.

Scudder, Samuel H. 1878. An Account of some Insects of unusual interest from the Tertiary Rocks of Colorado and Wyoming. [Description of the first American fossil Butterfly, *Prodryas persephone*]. *Bulletin of the United States Geological and Geographical Survey of the Territories,* 4(2): 519-543.

Scudder, Samuel H. 1889. The Fossil Butterflies of Florissant. Eighth Annual Report, U.S. Geological Survey. Pp. 439-474, pls. LII-LIII.

Scudder, Samuel H. 1890. The Tertiary Insects of North America. Report of the U.S. Geological Survey of the Territories, Volume XIII. Government Printing Office, Washington. 663 pp., 28 plates.

Scudder, Samuel H. 1892. Some insects of special interest from Florissant, Colorado, and other points in the Tertiaries of Colorado and Utah. *Bulletin of the United States Geological Survey,* No. 93: 1-25, 3 pls.

Shields, Oakley. 1976. Fossil butterflies and the evolution of Lepidoptera. *Journal of Research on the Lepidoptera,* 15(3): 132-143.

Shields, Oakley. 1985a. Southeast Asian affinities in Colorado Oligocene Libytheidae. *Tokurana Special,* No. 1 (Feb. 27, 1985): 13-23, Fig. 1-2.

Shields, Oakley. 1985b. Zoogeography of the Libytheidae (Snouts or Beaks). *Tokurana (Acta Rhopalocerologica),* No. 9 (April 14, 1985): 1-58, Fig. 1-60.

Shields, Oakley. 1987. The geologic significance of *Libythea collenettei* (Lepidoptera: Libytheidae) endemic to the Marquesas Islands, South-Central Pacific. *Bulletin of the Southern California Academy of Sciences,* 86(2): 107-112.

Shields, Oakley. 1988. Mesozoic history and neontology of Lepidoptera in relation to Trichoptera, Mecoptera, and Angiosperms. *Journal of Paleontology,* 62: 251-258.

Shields, Oakley. 1989a. Systematic position of Libytheidae, diphylogeny of Rhopalocera, and heteroceran ancestry of Rhopalocera (Lepidoptera). *Tyo to Ga,* 40: 197-228.

Shields, Oakley. 1989b. World numbers of butterflies. *Journal of the Lepidopterists' Society,* 43(3): 178-183.

Sperry, Grace H., and John L. Sperry. 1932. Notes on the larva of *Melitaea pola* Bdv. *Bulletin of the Southern California Academy of Sciences,* 31(1): 8.

Weber, William A. 1976. *Rocky Mountain Flora.* Colorado Associated University Press, Boulder, Colorado. 479 pp.

West, David A., and Wade N. Hazel. 1979. Natural pupation sites of swallowtail butterflies (Lepidoptera: Papilionidae): *Papilio polyxenes* Fabr., *P. glaucus* L., and *Battus philenor* (L.). *Ecological Entomology,* 4: 387-392.

GLOSSARY

ABDOMEN - the last or posterior major body division of an insect containing the digestive and reproductive organs.

ADULT - the last or reproductive stage of an insect.

AEDEAGUS - the penis or male reproductive organ.

ALPINE - the region of mountains above tree-line.

ANAL ANGLE - the angular portion of the hindwing nearest the end of the abdomen.

ANDROCONIA - the specialized scales with tufted tips that aid in the dissemination pheromones.

ANTENNA - the long, segmented appendage found on each side of the head between the eyes, which is thickened or clubbed at the tip in butterflies and thread-like or feathery in most moths.

APICULUS - the bent, tapering portion of the antennal club in skippers.

AUTHOR - the person who described and named a species or subspecies.

BASAL - toward the base or point of attachment nearest the body.

BASKING - behavior in which the body is fully exposed to the sun's radiation for warmth.

BEHAVIOR - anything an organism does involving action and/or response to stimulation.

BINOMIAL SYSTEM - method of designating organisms with a unique combination of two names, the genus and species.

BIVOLTINE - having two generations per year.

BROOD - a generation from egg to adult.

BUTTERFLY - a daytime-active insect having four scale-covered wings, sucking mouthparts, two reproductive openings in the female, and clubbed antennae.

CARDENOLIDES - a class of chemical compounds found primarily in plants belonging to the Scrophulariaceae, Apocynaceae, and Asclepiadaceae that cause cardiac arrhythmia, emesis, and visual disturbance to vertebrates. Example: Digitoxin. See CARDIAC GLYCOSIDE.

CARDIAC GLYCOSIDE - a non-volatile compound composed of a cardenolide and a glycoside (sugar). The usual form of cardenolides in plants.

CATERPILLAR - the fleshy, worm-like, feeding stage of moths and butterflies. See LARVA.

CENOZOIC ERA - the period of geological time extending from 65 million years ago to present. Commonly known as The Age of Mammals.

CHAPARRAL - a vegetation type that occurs in areas with hot dry summers and cool winters characterized by a dense growth of low evergreen shrubs.

CHRYSALIS - the pupa of a butterfly.

CLASPERS - part of the genitalia of male Lepidoptera used to hold onto the female during copulation.

CLASSIFICATION - the process of organizing species into groups usually based upon their geneology.

COCOON - a covering of silk spun by a fully-grown larva as protection for the pupa.

COMMON NAME - the name of an organism in general usage. See SCIENTIFIC NAME.

COMMUNITY - the ecological unit composed of all plant and animal populations living in a given area.

COMPOSITES - plants belonging to the Aster Family (Asteraceae).

COSTA - the thickened anterior margin of the forewings and hindwings.

COSTAL FOLD - a narrow flap covering a patch of androconial scales along the anterior margin of the forewings of some male skippers.

COURTSHIP - behavior to attract the opposite sex for mating.

CREMASTER - structure at the posterior tip of the pupa that usually bears hooked spines for attachment to the substrate.

CRYPTIC - colored or patterned to blend in with the surrounding environment.

DETOXIFYING ENZYME - proteins produced by caterpillars that break down toxins.

DIAPAUSE - a period of arrested development.

DICOTYLEDON - a class of flowering plants characterized by the presence of two seed leaves (cotyledons).

DISC - the central portion of the wing.

DISCAL - toward the center of the wing.

DIURNAL - active during the daytime.

DORSAL - the upper surface.

ECLOSION - the process whereby the larva hatches from the egg or the adult emerges from the pupa.

ECOLOGY - the study of the interrelationships between organisms and their environment.

EDAPHIC - resulting from or influenced by the soil.

EGG - the first stage of an insect.

ENDEMIC - restricted to a particular locality or region.

EPOCH - an interval of geologic time, shorter than a period or era.

EVOLUTION - the continuous genetic change of living organisms from generation to generation.

EXOSKELETON - the outer, supportive covering of insects.

FAMILY - a group of closely-related genera.

FLIGHT SEASON - the period of time during which adults are present.

FOODPLANT - a plant eaten by the larvae of a butterfly. See HOSTPLANT.

FOSSIL - the preserved remains or impressions of a previously-existing organism.

FRINGE SCALES - the scales projecting beyond the margin of the wing membrane.

GENITALIA - the organs and structures composing the reproductive system.

GENUS - a group of closely related species.

GREGARIOUS - living together in a group.

GROUND COLOR - the color occupying the largest area of the wing.

HABITAT - the kind of place where a particular organism lives.

HAIR PENCIL - a group of hair-like scales that are used by male butterflies to disseminate pheromones during courtship.

HERBARIUM - a collection of plant specimens.

HIBERNATION - a state of inactivity or dormancy.

HILLTOPPING - a type of mate-locating behavior in which males and unmated females frequent the tops of hills or trees.

HOLOTYPE - the single specimen selected by a describer of a species as its representative or type.

HOSTPLANT - a plant eaten by the larvae of a butterfly. See FOODPLANT.

IMMATURE - the egg, larval, or pupal stage.

INCIPIENT SPECIES - species in the process of coming into being.

INCREMENTAL GROWTH - the type of growth exhibited by invertebrates that must periodically shed (molt) the exoskeleton in order to become larger.

INDO-MALAYAN - the tropical areas of the Indian subcontinent and southeast Asia.

INFLORESCENCE - a flower-bearing stalk.

INSTAR - the various larval stages. The first instar is the stage between the egg and the first larval molt. Most butterflies have five larval instars.

INTERGRADE - hybridization between subspecies.

LAHARS - volcanic flows of water, mud, and rocky debris.

LARVA - the second or feeding stage of a butterfly. see CATERPILLAR.

LATERAL - toward the side.

LEPIDOPTERA - the insect order comprising moths and butterflies.

LIFE CYCLE - the series of stages during the life of a butterfly from egg to adult.

LONGITUDINAL - parallel to the body axis.

LUNULE - a crescent-shaped marking.

MEDIAN - the middle part of the wing.

METABOLISM - the sum of the various chemical activities going on within an organism.

METAMORPHOSIS - the series of changes through which an insect passes during its growth. Butterflies and moths have four life stages: egg, larva, pupa, and adult.

MIGRANT - a species which often disperses long distances.

MIOCENE - the period of geological time extending from 24.6 to 5.1 million years ago.

MOLT - to shed the exoskeleton.

MONOCOTYLEDON - a class of flowering plants characterized by the presence of only one seed leaf (cotyledon).

MONTANE - of the mountains.

MORTALITY - death rate.

MOTH - a usually nighttime-active insect having four scale-covered wings, one or two reproductive openings in the female, and thread-like or feathery antennae.

NEOTROPICAL - tropical areas of the Americas.

NOMENCLATURE - the process of giving names to organisms.

OCELLUS - an eye-like spot on the wings.

OLIGOCENE - the epoch of geological time extending from 38 to 24.6 million years ago.

ORDER - a group of closely-related families.

OSMETERIUM - the fleshy organ located behind the head of swallowtail caterpillars that can be everted when the larva is provoked. Strong-smelling and strong-tasting chemicals produced by the osmeterium deter predators.

OVERWINTER - to pass the climatic extremes of wintertime.

OVIPOSIT - to lay eggs.

PALEARCTIC - extratropical areas of Eurasia and northern Africa.

PALP - a three-segmented structure projecting in front of the face of adult moths and butterflies.

PARASITOID - an organism (usually a wasp or fly) whose immature stages feed on and eventually destroy the host (usually the egg, larva, or pupa of a butterfly or moth).

PATROLLING - a type of mate-locating behavior in which males actively search for females.

PERCHING - a type of mate-locating behavior in which males perch near areas frequented by females. Perching males fly out to investigate passing butterflies, some of which are receptive females.

PETIOLE - a stalk attaching a leaf to a stem.

PHENOTYPE - the appearance of an organism.

PHEROMONE - a chemical released by one individual that elicits a behavioral response in other individuals of the same species.

POSTMEDIAN - the area just beyond the middle of the wing.

PRE-CAMBRIAN ERA - the period of geological time extending from approximately 4.6 billion to 550 million years ago.

PREDATOR - an animal that consumes other organisms.

PROBOSCIS - the tongue-like sucking mouth parts of moths and butterflies.

PROTHORACIC SHIELD - a thickened area on the first part of the thorax of some caterpillars.

PUPA - the third stage of higher insects between the larval and adult stages.

RACE - a geographically distinct form of a species.

RANGE - the geographic area occupied by a species.

RAVINE - a small steep-sided valley that is larger than a gully and smaller than a canyon.

RESIDENT - locally breeding.

RIPARIAN - area along the banks of a natural watercourse.

SCALES - the flattened hairs covering the wings and bodies of moths and butterflies.

SCENT PAD - a cluster of androconial scales on the forewing of some male hairstreaks.

SCIENTIFIC NAME - a unique name consisting of two words, the genus and species, given to an organism; the scientific name is always italicized and is in Latin.

SEQUESTER - to hold, retain, or isolate, as in plant poisons sequestered by monarch larvae.

SKIPPER - a member of the butterfly families Hesperiidae or Megathymidae.

SPECIES - a group of organisms that is capable of interbreeding and is reproductively isolated from all other such groups.

SPECIES DIVERSITY - the number of species found in a particular area.

SPERMATOPHORE - a sac of sperm and accessory fluids produced by male butterflies during copulation.

SPHRAGIS - a light-weight plug secreted over the female vagina by certain male butterflies immediately after insemination.

SPIRACLE - the outer opening of the respiratory system in insects.

STIGMA - a patch of androconial scales located in the middle of the forewings of some male skippers.

STRAY - an individual far beyond the normal range of the species.

SUBFAMILY - within a family, a group of closely-related genera.

SUBMARGINAL - refers to the portion of the wing just before the outer margin.

SUBSPECIES - a distinctive geographical race of a species.

SUBTROPICAL - at the boundaries of the tropical and temperate zones.

SUPERFAMILY - a group of closely-related families.

SURANAL PLATE - a thickened area located on the upper posterior end of some caterpillars.

SYSTEMATICS - the classification and study of organisms with regard to their evolutionary relationships.

TAXONOMIC CATEGORIES - the units of classification which for all animals minimally consist of kingdom, phylum, class, order, family, genus, and species.

TAXONOMY - the science of classifying organisms.

THECLA SPOT - an eyespot at the anal angle of the hindwings in many lycaenid butterflies.

THORAX - the middle portion of an insect to which the legs and wings are attached.

TIBIA - the fourth division of the leg located between the femur and the tarsus.

TOXIN - a poisonous substance.

TRANSVERSE - lying at a right angle to the body axis.

TRINOMEN - a scientific name consisting of the genus, species, and subspecies.

TROPICAL - that portion of the earth lying between the Tropic of Cancer and the Tropic of Capricorn that does not experience freezing temperatures.

TUBERCLE - a projection on a caterpillar or pupa.

TYPE LOCALITY - the geographic area from which the holotype of a species was collected.

UNIVOLTINE - having a single generation each year.

UNPALATABLE - distasteful or unpleasant.

VEINS - the rod-like structures supporting the wings of insects.

VENTRAL - the under side.

WARNING COLORATION - brilliant or conspicuous coloration used to advertise noxious qualities in particular organisms.

CHECKLIST OF THE PRESENT-DAY BUTTERFLIES OF THE FLORISSANT REGION OF CENTRAL COLORADO

HESPERIIDAE
Subfamily PYRGINAE

_____ 1. *Thorybes mexicana nevada* Scudder
_____ 2. *Erynnis icelus* (Scudder and Burgess)
_____ 3. *Erynnis funeralis* (Scudder and Burgess)
_____ 4. *Erynnis afranius* (Lintner)
_____ 5. *Erynnis persius fredericki* H. A. Freeman
_____ 6. *Pyrgus centaureae loki* Evans
_____ 7. *Pyrgus xanthus* W. H. Edwards
_____ 8. *Pyrgus scriptura* (Boisduval)
_____ 9. *Pyrgus communis* (Grote)

Subfamily HESPERIINAE

_____10. *Oarisma garita* (Reakirt)
_____11. *Stinga morrisoni* (W. H. Edwards)
_____12. *Hesperia uncas uncas* W. H. Edwards
_____13. *Hesperia comma ochracea* Lindsey
_____14. *Hesperia nevada* (Scudder)
_____15. *Polites draco* (W. H. Edwards)
_____16. *Polites themistocles* (Latreille)
_____17. *Polites sonora utahensis* (Skinner)
_____18. *Atalopedes campestris huron* (W. H. Edwards)
_____19. *Ochlodes snowi* (W. H. Edwards)
_____20. *Euphyes ruricola metacomet* (Harris)

PAPILIONIDAE
Subfamily PARNASSIINAE

_____21. *Parnassius phoebus sayii* W. H. Edwards

Subfamily PAPILIONINAE

_____22. *Papilio polyxenes asterius* Stoll
_____23. *Papilio zelicaon nitra* W. H. Edwards
_____24. *Papilio rutulus rutulus* Lucas
_____25. *Papilio multicaudatus* Kirby
_____26. *Papilio eurymedon* Lucas
_____27. *Battus philenor philenor* (Linnaeus)

PIERIDAE
Subfamily PIERINAE

_____28. *Neophasia menapia* (C. and R. Felder)
_____29. *Pontia sisymbrii elivata* (Barnes and Benjamin)
_____30. *Pontia protodice* (Boisduval and Leconte)

_____31. *Pontia occidentalis occidentalis* (Reakirt)
_____32. *Pieris (Artogeia) rapae* (Linnaeus)

Subfamily ANTHOCHARINAE

_____33. *Euchloe ausonides coloradensis* (Henry Edwards)

Subfamily COLIADINAE

_____34. *Colias philodice eriphyle* W. H. Edwards
_____35. *Colias eurytheme* Boisduval
_____36. *Colias alexandra alexandra* W. H. Edwards
_____37. *Eurema nicippe* (Cramer)
_____38. *Eurema mexicana* (Boisduval)
_____39. *Nathalis iole* Boisduval

LYCAENIDAE

Subfamily LYCAENINAE

_____40. *Lycaena (Chalceria) rubidus sirius* (W. H. Edwards)
_____41. *Lycaena (Chalceria) heteronea heteronea* (Boisduval)
_____42. *Lycaena (Epidemia) helloides* (Boisduval)

Subfamily EUMAENINAE

_____43. *Harkenclenus titus titus* (Fabricus)
_____44. *Satyrium behrii crossi* (Field)
_____45. *Satyrium californica* (W. H. Edwards)
_____46. *Satyrium calanus godarti* (Field)
_____47. *Callophrys apama homoperplexa* Barnes and Benjamin
_____48. *Mitoura spinetorum* (Hewitson)
_____49. *Mitoura siva siva* (W. H. Edwards)
_____50. *Incisalia polios obscurus* Ferris and Fisher
_____51. *Incisalia eryphon eryphon* (Boisduval)
_____52. *Strymon melinus franki* Field

Subfamily POLYOMMATINAE

_____53. *Hemiargus isola alce* (W. H. Edwards)
_____54. *Everes amyntula amyntula* (Boisduval)
_____55. *Celastrina ladon cinerea* (W. H. Edwards)
_____56. *Glaucopsyche lygdamus oro* Scudder
_____57. *Lycaeides melissa melissa* (W. H. Edwards)
_____58. *Plebejus saepiolus whitmeri* F. M. Brown
_____59. *Agriades franklinii rustica* (W. H. Edwards)

RIODINIDAE

_____60. *Apodemia mormo duryi* (W. H. Edwards)

NYMPHALIDAE

Subfamily HELICONIINAE

_____61. *Agraulis vanillae incarnata* (Riley)

Subfamily NYMPHALINAE

_____62. *Polygonia interrogationis* (Fabricus)
_____63. *Polygonia satyrus* (W. H. Edwards)
_____64. *Polygonia faunus hylas* (W. H. Edwards)
_____65. *Polygonia zephyrus* (W. H. Edwards)
_____66. *Nymphalis (Nymphalis) antiopa antiopa* (Linnaeus)
_____67. *Nymphalis (Aglais) milberti milberti* Godart
_____68. *Vanessa virginiensis* (Drury)
_____69. *Vanessa cardui* (Linnaeus)
_____70. *Vanessa atalanta rubria* (Fruhstorfer)
_____71. *Euptoieta claudia* (Cramer)
_____72. *Speyeria idalia* (Drury)
_____73. *Speyeria edwardsii* (Reakirt)
_____74. *Speyeria coronis halcyone* (W. H. Edwards)
_____75. *Speyeria atlantis electa* (W. H. Edwards)
or *hesperis* (W. H. Edwards)
_____76. *Speyeria mormonia eurynome* (W. H. Edwards)
_____77. *Phyciodes tharos pascoensis* Wright
_____78. *Phyciodes campestris camillus* W. H. Edwards
_____79. *Phyciodes vesta* (W. H. Edwards)
_____80. *Phyciodes pallidus* (W. H. Edwards)
_____81. *Chlosyne (Charidryas) gorgone carlota* (Reakirt)
_____82. *Chlosyne (Charidryas) palla calydon* (Holland)
_____83. *Chlosyne (Thessalia) fulvia* (W. H. Edwards)
_____84. *Poladryas arachne arachne* (W. H. Edwards)
_____85. *Euphydryas (Occidryas) anicia capella* (Barnes)
_____86. *Limenitis (Basilarchia) weidemeyerii weidemeyerii* (W. H. Edwards)

Subfamily SATYRINAE

_____87. *Cyllopsis pertepida dorothea* (Nabokov)
_____88. *Coenonympha tullia ochracea* W. H. Edwards
_____89. *Cercyonis meadii meadii* (W. H. Edwards)
_____90. *Cercyonis oetus charon* (W. H. Edwards)
_____91. *Erebia epipsodea epipsodea* Butler
_____92. *Neominois ridingsii ridingsii* (W. H. Edwards)
_____93. *Oeneis chryxus chryxus* (Doubleday and Hewitson)
_____94. *Oeneis uhleri uhleri* (Reakirt)
_____95. *Oeneis alberta oslari* Skinner

Subfamily DANAINAE

_____96. *Danaus plexippus plexippus* (Linnaeus)
_____97. *Danaus gilippus strigosus* (Bates)

CHECKLIST OF THE FOSSIL BUTTERFLIES KNOWN FROM THE FLORISSANT FOSSIL BEDS

PIERIDAE
Subfamily PIERINAE

______ 1. *Stolopsyche libytheoides* Scudder
______ 2. *Oligodonta florissantensis* Brown

LIBYTHEIDAE
Subfamily Libytheinae

______ 3. *Prolibythea vagabunda* Scudder
______ 4. *Barbarothea florissanti* Scudder

NYMPHALIDAE
Subfamily NYMPHALINAE

______ 5. *Prodryas persephone* Scudder
______ 6. *Lithopsyche styx* Scudder
______ 7. *Jupiteria charon* Scudder
______ 8. *Nymphalites obscurum* Scudder
______ 9. *Nymphalites scudderi* Beutenmuller and Cockerell
______10. *Apanthesis leuce* Scudder
______11. *Chlorippe wilmattae* Cockerell
______12. *Vanessa amerindica* Miller and Brown

PHOTO CREDITS FOR TEXT FIGURES

Front cover - Boyce A. Drummond
1 - Maria F. Minno
2 - (Map) Daryl Harrison
3 - Thomas C. Emmel
4 - Thomas C. Emmel & F. Martin Brown
5 - Boyce A. Drummond
6 - Marc C. Minno
7 - Marc C. Minno
8 - Boyce A. Drummond
9 - Boyce A. Drummond
10 - Boyce A. Drummond
11 - Boyce A. Drummond
12 - Marc C. Minno
13 - Marc C. Minno
14 - Thomas C. Emmel
15 - Boyce A. Drummond
16 - Marc C. Minno
17 - Marc C. Minno
18 - Boyce A. Drummond
19 - Marc C. Minno
20 - Marc C. Minno
21 - (Above) - Marc C. Minno
22 - (Below) - Boyce A. Drummond
22 - Marc C. Minno
23 - Boyce A. Drummond
24 - Marc C. Minno
25 - Marc C. Minno
26 - Marc C. Minno
27 - Marc C. Minno
28 - Marc C. Minno
29 - Marc C. Minno
30 - (Left) - Marc C. Minno
30 - (Right) - Boyce A. Drummond
31 - Marc C. Minno
32 - Marc C. Minno
33 - Thomas C. Emmel
34 - Marc C. Minno
35 - Marc C. Minno
36 - Marc C. Minno
37 - Marc C. Minno
38 - Marc C. Minno
39 - Marc C. Minno
40 - Marc C. Minno
41 - Marc C. Minno
42 - Marc C. Minno
43 - Marc C. Minno
44 - Marc C. Minno
45 - Marc C. Minno
46 - Marc C. Minno
47 - Thomas C. Emmel
48 - Thomas C. Emmel
49 - Marc C. Minno
50 - Boyce A. Drummond
51 - John F. Emmel
52 - Boyce A. Drummond
53 - Marc C. Minno

INDEX TO PLANT SPECIES

GENERAL INDEX

Printed and bound by CPI Group (UK) Ltd, Croydon, CR0 4YY

16/04/2025

14658408-0001